REDISCOVERING GANDHI

VOLUME 5 : Humanizing Science and Technology: Some Case Studies

Series Editor: ***Prof. R.P. Misra***

Vol. 1. Hind Swaraj: *Gandhi's Challenge to Modern Civilization*
—R.P. Misra

2. Ahimsa: *The Way to Peace*
—Jonathan K. Crane & Jordi Augusti-Panareda

3. Satyagraha: *Gandhian Approach to Conflict Resolution*
—Jainarain Sharma

4. Consensual Democracy: *Gandhi on State, Power and Politics*
—Ashu Pasricha

5. Humanizing Science and Technology: *Some Case Studies*
—Vinnie Jauhari and Parul Wasan

6. Gandhi and Jesus: The End of Fundamentalism
—Wolfgang Sternstein

ABOUT THE AUTHORS

Dr. Vinnie Jauhari, is M.Sc.(Hons.) in Electronics from Panjab University, Chandigarh and MBA in Marketing. She received her Ph.D. on Corporate Entrepreneurship from IITD and her Post-Doc from the United Nations University, Tokyo, in the area of Technology and Society. She is currently Director, IIMT, Gurgaon. She was earlier heading Open Innovation Office for HP Labs India. She is also the founding editor of the Journal of Services Research, has over 100 publications in National and International Journals and has also authored 8 books on Business Strategy and Services Management. Her area of expertise is Corporate Entrepreneurship, Strategy and Technology Management.

Parul Wasan (Ph.D.), has taught Business Communications, Personal Skill Profiling and Placement Skills, and Interactive Behaviour at Work, at the Institute for International Management and Technology, Oxford Brookes University, Gurgaon, Haryana, India. She has also been the editorial coordinator for the Journal of Services Management (an international journal listed on EBSCO data base). She has to her credit number of academic papers which have been published in reputed journals. She is a Gold Medalist in Communication in English. She has spent her formative years in media and television production.

REDISCOVERING GANDHI

Series Editor
Prof. R.P. Misra

Volume 5
Humanizing Science and Technology: Some Case Studies

Vinnie Jauhari and **Parul Wasan**

Published for
Centre for Gandhi Peace Studies
Indira Gandhi National Open University, New Delhi
by
CONCEPT PUBLISHING COMPANY PVT. LTD.
NEW DELHI-110059

ISBN-13: 978-81-8069-374-8 (Series)
ISBN-13: 978-81-8069-776-0 (Vol. V)

First Published 2014

Published and Printed by

Concept
Since 1974

Concept Publishing Company Pvt. Ltd.
A/15-16, Commercial Block, Mohan Garden
New Delhi-110059 (India)
T: 25351460, 25351794, *F*: 091-11-25357109
E: publishing@conceptpub.com, W: www.conceptpub.com
Editorial Office: H-13, Bali Nagar, New Delhi-110 015, India.

Cataloging in Publication Data--*Courtesy:* D.K. Agencies (P) Ltd. <docinfo@dkagencies.com>

Jauhari, Vinnie.
Humanizing science and technology : some case studies / Vinnie Jauhari and Parul Wasan.
p. cm. — (Rediscovering Gandhi ; v. 5)
Includes bibliographical references and index.
ISBN 9788180697760
ISBN 9788180693748 (series)

1. Science—Social aspects—India. 2. Technology—Social aspects—India. 3. Science—Economic aspects—India. 4. Technology—Economic aspects—India. 5. Sustainable development—India. 6. Social responsibility of business—India. 7. Gandhi, Mahatma, 1869-1948—Philosophy. I. Wasan, Parul, joint author. II. Title. III. Series: Rediscovering Gandhi ; v. 5.

DDC 303.4830954 23

Dedicated to my son
Shaurya Jauhari

Vinnie Jauhari

Dedicated to my daughter
Rhea Nilanjana

Parul Wasan

Editor's Foreword

Was Gandhi against science and technology? This question has significance because there is general misinformation about Gandhi's attitude in this regard. Term technology was not in currency in Gandhi's times. He used the term machine which in his vocabulary stood for technology. Gandhi was not against technology, although at times he appeared to be so. When seen in the total context of his philosophy of life and all that he has said on machine and its use, it is clear that he was not against machine *per se*.

Let us look at Gandhi's responses to questions raised on this issue in 1924.

> "*How can I be (against machinery) when I know that even this body is a most delicate piece of machinery? The spinning wheel is a machine; a little toothpick is a machine. What I object to is the craze for machinery, not machinery as such. The craze is for what they call labour-saving machinery. Men go on 'saving labour' till thousands are without work and thrown on the open streets to die of starvation. I want to save time and labour not for a fraction of mankind but for all. Today machinery merely helps a few to rid on the backs of millions. The impetus behind it all is not the philanthropy to save labour, but greed. It is against this constitution of things that I am fighting with all my might. . . . the supreme consideration is man. The machine should not tend to atrophy the limbs of man. For instance, I would make intelligent exceptions. Take the case of the Singer's sewing machine. It is one of the few useful things ever invented, and there is a romance about the device itself.*"
>
> "*There would have to be factory for making these sewing machines, and it would have to be power-driven machinery of ordinary type.*" asked the questioner. "*Yes*', Gandhi replied, "*but I*

am socialist enough to say that such factories should be nationalized, state controlled... the savings of the labour of the individual should be the object and not human greed the motive. Thus, for instance, I would welcome any day machines to straighten crooked spindles. Not that blacksmiths will cease to make spindles; they will continue to provide spindles but when the spindle goes wrong every spinner will have a machine to get it straight. Therefore, replace greed by need and everything will be all right."

"*But if you make an exception of the Singer's sewing machine and your spindle where would these exceptions end?*" asked the questioner. "*Just where they cease to help the individual and encroach upon his individuality*", said Gandhi. "*The machine should not be allowed to cripple the limbs of man.*" To the question "*But ideally, would you not rule out all machinery? When you except the sewing machine, you will have to make exceptions of the bicycle, the motor car etc.*" "*No, I don't*" replied Gandhi "*because they do not satisfy any of the primary wants of man; for it is not the primary need of man to traverse distances with rapidity of a motor car. The needle on the contrary happens to be an essential thing in life, a primary need.*" And he added, "*Ideally, I would rule out all machinery, even as I would reject this very body, which is not helpful to salvation, and seek the absolute liberation of the soul. From that point of view I would reject all machinery, but machines will remain, like the body, they are inevitable. The body itself, as I told you, is the purest piece of mechanism; but if it is a hindrance to the highest flights of the soul, it has to be rejected.*"[1]

The above quotations are enough to prove that the main issue before Gandhi was not machinery; it was the question of employment and gainful engagement for all. On a higher plane, it was the question of human survival and development. For him, work was worship and 'an idle mind is a devil's workshop.' Gandhi's objective of life was to raise the standard of life but it could not be done without raising the standard of living to a level a civilized human being deserves. The ultimate objective of Gandhi was help people move towards self-realization. Anything that came in the way of this goal must be shunned, machinery included. To create a society where peace and harmony reign supreme, all must share the gifts of nature as equitably as possible. Tendency to usurp power by a few and to deprive others of their legitimate share is immoral and unethical. Machinery helps concentrate power and resources in fewer hands; it, therefore, deserves to be kept under control.

A rich man finds it difficult to enter the Kingdom of God (soul) because his mind is set on money and all that it can buy. For Gandhi, the material and spiritual aspects of life were not juxtaposed; they were two sides of the same coin. One does not have to renounce the world to realize the self. It is not World *versus* Soul; Matter *versus* Spirit; Prakriti *versus* Purush; and Pleasure *versus* Happiness; it is World *circumscribed by* Soul *in* Matter *circumscribed by* Spirit; Prakriti *circumscribed by* Purush; and Pleasure *circumscribed by* Happiness. We don't have to reject either, for both are essential parts of our life. Applying to our own individual selves, we have body that is Matter in which resides the Soul. The Soul without body would merge in the Super Soul (*Parmatma* = God = Zero). Matter without Spirit would lead to violence; Prakriti without Purush would end in global warming; and Pleasure without Happiness would be suicidal at best.

If machinery can help improve the quality of life of the people and remove poverty and pestilence, it should be welcome for bread is the God of the hungry. If it can empower people and make them stand on their own feet, it must be welcome. But if it is used to amass wealth, enslave and kill people, and destroy nature, it must be shunned. Machinery can thus become a source of bliss if it is created and used by those who look for God in everything around them; who look for Spirit in Matter; Purush in Prakriti; and Happiness in Pleasure. But if it is used to enslave others or deprive people of their legitimate rights, it would lead to unhappiness. Gandhiji has himself said that he was not against the use of tools and machines, simple or sophisticated if they helped the poor and downtrodden get employment. He disliked machines that rendered people unemployed. So while opposing machines, he himself used them, and thus pointed out that the application and use of machinery renders it good or bad not the very contrivance but by the purpose for which it is used. Machine is a means not an end and it has to be used for good causes and objectives.

It is unfortunate that man has come to believe that all that God has created is his, and he should defend his right to control the world by all the means at his disposal. If he is not able to do it, it is because there are other claimants with whom he has to contend with. Then he joins hands with those having similar objectives and tries to share the resources of the earth with them. And thus the society is structured to have haves like landlords, industrialists, businessmen, and mafias and have-nots like small farmers, landless agricultural workers, and beggars. But even the haves are not happy. Fighting among them is the order of the day as everyone wants to enlarge his power but none wants to share what he has with others. Everyone wants to have more than what others have. And thus there is no peace.

All over India people are agitating to save their livelihood. Multinational and national corporations are depriving them of their land leaving them in wilderness to fend for themselves. Their eyes are fixed on the agricultural land, on the forests, on water bodies, and on mineral rich areas. they are vying with each other to enlarge their possessions using all the physical force at their command. Machines, the byproducts of technology put tremendous power in the hands of the contending parties. All weapons of mass destruction are the products of technology. Misuse of technology can lead to mass killing of humanity and mass destruction of the biotic world. Violence that plagues humanity today becomes more devastating when technology comes to its aid. The global warming and climate change that stares at us today has also resulted from the misuse of technology.

Thus we see that Gandhi was right in saying that technology in itself is not dangerous; it becomes dangerous when it is put in wrong hands. And wrong hands are those that seek pleasure instead of happiness, conflict instead of harmony, hatred instead of love, and war instead of peace. Thus, Gandhi was not against technology; he was against the misuse of technology. Anything that came in the way of man becoming a better human being was shunned by Gandhi for he knew that only good human beings can create a society that is egalitarian, decentralized, peace loving, and progressive.

In 'Hind Swaraj' (1909) Gandhi attacked the western model of development based as it is on the supremacy of machines, and told bluntly that: "*Machinery has begun to desolate Europe. Ruination is knocking at the English gates.*" "*Machinery is the chief symbol of modern civilization; it represents a great sin*" and "*Machinery is like a snake-hole which may contain from one to a hundred snakes. Where there is machinery there are large cities; and where there are large cities, there are tram cars and railways; and there only does one see electric light. English villages do not boast any of these things……I cannot recall a single good point in connection with machinery. Books can be written to demonstrate its evils.*"[2]

Partially modifying his views on machines, he said in mid-1920s,

"I have nothing to withdraw from what I said about machines in the Indian Home Rule……I do not suggest today a destruction of all the machines, but I am making spinning wheel the master machine. The Indian Home Rule depicts an ideal state'.[3] .... And added "*I had no quarrel with rail roads, steamers and many machines as such, but I protested against the abuse that was at present being made of them, either for exploiting many nations of the earth or for destroying them*".[4] He became still more positive when he said "*Machinery has its place, it has come to stay. But it must not be allowed to displace the necessary human labour……I would welcome every improvement in the cottage machine, but I know that it is criminal to displace the hand labour by the introduction of power driven spindles unless one is, at the same time, ready to give millions of farmers some other occupation in their homes*".[5]

He was not avers to even importing machines if a situation so demands. Look at what he said with regard to the use of hides of the cattle. "*As long as we cannot make the machine required for utilizing the hide of dead cattle, worth nine crores, available in our country, I would be ready to import them from any part of the world*".[6] Similarly, in an interview to Edmond Demeter in 1931, he clarifies, "*I am an enemy not of mechanism but of organized mechanism. I consider this system, which has become the basis of your civilization, as the greatest danger which could menace man*".[7]

Who would not agree with the Mahatma when he says: "*Mechanization is good when hands are too few for the work intended to be accomplished……It is an evil when there are more hands than required for the work, as is the case in India*".[8] As his views became less acceptable to his followers, he conceded that "*I do visualize electricity, ship-building, iron works, machine making and the like existing side by side with village handicrafts. But the order of dependence will be reversed. Hitherto industrialization has been so planned as to destroy the villages and the village crafts. In the state of future it will subserve the villagers and their crafts*".[9]

Gandhi Ji did not think of technology as having just one relevance i.e. production of goods and services. It was integrated in his political, economic and social thinking as well. To him choice of technology had its implications for democracy, community development, progress, employment, equity, education, health, production and distribution. He did not rule out large scale industries. *In his own words: "The movement of spinning wheel is an organized attempt to displace machinery from that state of exclusiveness and exploitation and to place it in its proper state. Under my scheme, therefore, men in charge of machinery will think not of themselves or even of the nations to which they belong but of the whole human race".*[10]

You cannot build non-violence on a factory civilization, but it can be built on self-contained villages. Even if Hitler so minded, he could not devastate seven hundred thousand non-violent villages. He would himself become non-violent in the process. Rural economy as I have conceived eschews exploitation altogether and exploitation is the essence of violence. You have, therefore, to be rural minded before you can be non-violent, and to be rural minded, you have to have faith in the spinning wheel.[11]

Gandhi did concede that "*Large scale industries are needed in exploiting natural resources which should form the raw materials of the people's industries. Control of such raw materials cannot be left in private hands. Therefore, all such industries should also be under the state. In economics, large scale industry is the antithesis of democracy in politics. It is not by chance that the western nations have come by their economic organization. It is a result of their way of thinking in terms of autocracy. They find themselves with dictatorships in political organization, and centralized industries in the economic field. These two go together and we cannot have the one without the other.*

"*If we aim at democracy, we have to abandon centralized production in consumption goods. This does not preclude all centralized industries. As we have already seen there is a definite field of such, though restricted*".[12]

Elaborating further, he said in 1936: "*There is nothing good or evil in inanimate things. These attributes come into existence by man's reaction to these. A thing is good or evil according to as we put to benefit or harmful use...... Machines can be masters as well as servants......Machine or no machine, is not the fundamental question. The problem is one of finding the best way of not only supplying the material wants but also bringing employment to the people. Machinery which only saves labour is a curse in an unemployed or underemployed India, whatever it may be in other countries. Machinery can only be used to make such things as hands cannot make and where will not displace workers gainfully employed.*[13]

Knowledge and technology are conceived to be common properties and in case of large industries, it was argued by him that it should be state owned. "*In the village movement there is a definite place for centralized industries, not for their own sake, but as adjuncts and subsidiaries to decentralized units, just as in the political sphere a democracy does not eschew governmental control and regulations when it is directed towards the better realization of individual development and expression.*"[14]

Gandhi tested the benignity of machinery or technology or even knowledge if it affirmatively answered the following questions :

1. Does it help man to become a better human being?
2. Does it empower individuals and communities to become self-reliant and promotes development from below?
3. Does it help create full employment?
4. Does it ensure democratic participation of all?
5. Does it help nature to nurture life on planet earth?

If the answer to all these questions is yes, Gandhi would stand like a rock for it; but if the answer is no, Gandhi would oppose and shun it. So, despite contradictory statements made by Gandhi on machinery in different context, we can conclude that he was not against machinery *per se*. He was against machinery if it were used to enslave people, to render them unemployed, to centralize power in the hands of a few, and to weaken grassroots democracy. He advocated the use of simple machines that the common man living even in remote villages could use to earn his livelihood and meet his basic needs. To say that he wanted humanity to go back to the stage of noble savage is to twist his statements and deliberately deny the truth underlying them.

It is unfortunate that Gandhi's ideas in this respect were rejected outright even in India after 1947 and in the new millennium they are considered obsolete. During the freedom struggle, Gandhi was the undisputed leader of the mainstream freedom fighters. Wherever he went, thousands followed him. He was undisputed and indispensable leader of the masses. He lived like a peasant, moved like a sportsman, talked like a friend, and worked like a mason building a new society brick by brick. He knew how the edifice would look like but before he could complete the foundation of new India he imagined, he fell to the bullets of a misguided youth and met the fate of all those who tried to change the world for better but found the task too big for any single individual, no matter how great he might be.

Once he was off the scene, his popularity among the leaders of the freedom movement waned by the time India became a Sovereign Republic. Industrialization and urbanization gave way to Gandhi's handcrafts and village republics. Did the west not develop because of industrialization and urbanization? Did the USSR not become a modern nation by industrializing and urbanizing itself? Gandhi would keep India steeped in village crafts, peasant farming, poverty, and pestilence, felt the leaders who took the reigns of government once the British left the country.

India became an independent nation on August 15, 1947. This great event triggered a call for independence all over the world and country after country gained independence in Asia and Africa. But India could not generate a new wave of development that was different from what the west, the tormenter of the rest of the world, was pursuing. India remained tied to the west intellectually and economically. It inherited all the institutions the British had built to rule over India. Leaving aside some sugar coatings here and there, these institutions continue to hold their sway even after six decades of independence. Indian history was written by the western scholars with a biased mind. Gandhi told repeatedly that true history of a people can never be written by foreigners. But our children continue to mug the history of India as written by foreigners. English continues to hold a pride of place, the richness of Indian languages all the way from Sanskrit to Tamil notwithstanding. We feel pride in using foreign goods, fashions, and styles of life and in deriding our own cultural heritages as antiquated, backward and unmodern. India thus began its independent journey with the same institutions, development goals, and cultural ethos that the west had. The British left India but the British ways were too deeply ingrained in our minds to leave. India continues to be ruled by the west in Indian garb.

Gandhi was soon consigned to Rajghat and a few *ashrams* and institutes which organize functions on his birthday (October 2) and death anniversary (January 30). He became a subject matter of books written by foreigners and Indians; and seminars and conferences organized and attended mostly by those who had read his writings but never put his thoughts into practice. Gandhi's greatness as a thinker lay in the fact that his ideas emerged from practice and experiment and, therefore he went on changing them as he made new discoveries. There is not a single instance of the Government of India selecting a district or even a village to experiment with the

Gandhian model of development with right mixture of agriculture, small-scale industries, grassroots democracy and social regeneration. Even the Panchayatiraj institutions created in 1993 have been usurped by the politicians and bureaucrats; they have become tools in the hands of the government officials to siphon funds meant for social welfare and rural development.

Today, there is a growing craze for modern technology unmindful of its implications for people, especially the poor, and for ecology and environment. After more than six decades of planned development poverty remains our problem number 1. While various government agencies quarrel about how many poor are there in India and at what rate is poverty declining or growing, the fact remains that there are as many poor in India today (> 300 million) as was the total population of India (350 million) in 1947. To say that the population has grown over three times since then and, therefore, the number of the poor has also gone up amounts to admitting that we have failed in our mission of poverty eradication that Gandhi put in the forefront of the freedom struggle, during the last six decades. We are glad to know that number of billionaire and millionaire Indians is increasing fast and that we are not much behind the developed parts of the world in this respect. But this happiness is marred when we realize that millions of our brethren who die prematurely because of under—and malnutrition is also increasing. Thirty billionaires of the USA have pledged to donate half of their assets (50 per cent) for charity; but not a single Indian billionaire has so far come forward to set even 5 per cent of his assets apart for charitable purposes! If all the ill-gotten money deposited in Swiss Banks is brought back to India, it will be more than enough to build enough good schools and recruit teachers to educate children and poverty will be gone in no time. This is the state of affairs in the land where Gandhi lived and advocated trusteeship for all those who have more than what they need to live comfortably.

Think of a situation in which each village community becomes a socio-territorial unit working in cohesion to meet its basic needs; and city neighbourhoods managing their own local affairs? Think also of a pattern of human settlements wherein villages and cities constitute a common whole all linked together vertically as well as horizontally and producing goods and services they are most suited for. Instead, we have made the mega cities as engines of growth and villages as engines of decay. In between are the intermediate and small size towns and cities trying hard to sustain themselves. The pattern of urbanization we are promoting is leading to disasters of various kinds: natural, economic, and social. The cities are sucking the physical as well as the human resources of the countryside; villages are becoming dormitories where the labour working in cities resides. If the present rate of distress migration from rural to urban areas continues, villages will become desolate places and cities will be so overpopulated, polluted and violence prone that no one will be safe there. India is going to face a major problem of urban obesity and rural decay by 2025.

And add to all this, the all embracing problem of global warming and climate change. This problem is the direct consequence of technology we have adopted to develop our economy. We are robbing the nature of its capacity to regenerate itself and keep fit for human survival and development. Efforts are on at the global as well as national levels to tackle this problem, but no one wants to go to the roots of the problems. The question being posed is "How to reduce the CO_2 in the atmosphere and who should do it and to what extent?" and not "How to stop the release of CO_2 in the atmosphere by making our development style green?" And we know that we are not going to succeed in this endeavour because the way we are developing our economy and technology is counterproductive. But all is not lost. The lamp Gandhi lighted in his life time is not yet gone. A number of individuals and institutions have taken up the task of keeping the light on. They have proved that technology can be humanized; that it can make the life of the poor even in the midst of modernization and globalization sweeter; and that the power of the people is far stronger than the power of the machine. If this power of the people can become the central theme around which our development policy is woven, the results would be far better than what we see today. We would not only increase our GDP but also our GNH. Our democracy would be far more vibrant and people centric and abject poverty in which at least one-third of the people of India live today would become a thing of the past.

This volume authored by Dr. Vinnie Jauhari and Parul Goyal have put together some of the experiments being conducted in India to involve the poorer sections of the society in their own development. It examines the role of science and technology in creating an egalitarian and sustainable socio-economic order. There is a need to balance the development of economy, technology, and social welfare. Development must be people centric; anything that marginalizes some people constitutes maldevelopment.

The volume tells vividly that Mahatma Gandhi, despite our efforts to marginalize him has remained centre stage in the development debate in India as well as abroad. The case studies point to the success of the Gandhian model of decentralized and participatory development. There are many other cases in India like the AMUL (cooperative dairying) in Gujarat which have proved to be more efficient than large industrial enterprises. And case of lunch in *Dibbas* distributed to thousands of office employees daily on time and maintaining quality is unique to Mumbai, the financial capital of India. No modern restaurant can ever achieve this feet. It is possible to make a group of five to ten villages as an agro-industrial unit that produces and processes foodgrains, vegetables, fruits, oil seeds, and other commercial crops.

Let us not forget the fact that modern technology is not labour intensive. It suites well the EU and Anglo American countries for here population growth has almost come to a standstill. Countries like India and China have a backlog of huge population to take care and this population is still growing rather fast. These countries, for a long time to come, need technologies that are labour intensive along with those that are labour saving. A right mixture of the labour saving major industrial enterprises and labour intensive small and intermediate size units is the need of the hour.

Let us not think of promotion of small and intermediate size units as stop-gap arrangement. It should be treated as a long-term arrangement as the ecological consequences of large scale industrial units are already highly devastating. As time passes, environmental problems are going to further exasperate and the day is not far off when the use of fossil fuels will decrease rather abruptly and the world would return to green technology. Countries which have a mixture of green and red technologies would be better placed to switch over to small and intermediate size green machines and enterprises.

Apart from what Gandhi said and did, the emerging future demands the promotion of green technologies and development from below. The case studies presented in this book point to the potentials of production and distribution by popular participation on mass scales. Let us hope that the books like this would encourage people in different parts of India to launch similar ventures enlivening the life of millions of our countrymen who are un- or under-employed and who suffer from the heavy burden of poverty and pestilence. If it happens, India would be able to demonstrate to the world that there is another way of development that is more humane and people-centric. And it is the Gandhian Way.

Let me take this opportunity to thank the Centre for Gandhi and Peace Studies, Indira Gandhi National Open University (IGNOU), New Delhi for sponsoring the publication of the series (volume V) onwards especially its coordinator Prof. D. Gopal and Shri Ashok Kumar Mittal, Concept Publishing Company Pvt. Ltd. for publishing the series. I am quite sure the volume would be as popular as the earlier ones.

R.P. Misra
Series Editor

REFERENCES

1. Homer Alexander Jack (ed.), *The Gandhi Reader: A Sourcebook of His Life and Writings*, New York: Grove Press, 1956.
2. Gandhi, M.K., (1924) *Hind Swaraj,* Ahmedabad: Navajivan, Chapter 19.
3. *CWMG*, Vol. 28, p. 548.
4. Gandhi, M.K. *Young India* 18-8-1925, see *CWMG*, Vol. 28, p. 63.
5. Gandhi, M.K., *Young India* 5-11-1925, see *CWMG*, Vol. 28, p. 428.
6. Gandhi, M.K., (1927) *Navajivan*, see *CWMG.*
7. Gandhi, M.K., (1931) Interview to Edmond Demeter see *CWMG*, Vol. 48, p. 386.
8. Gandhi, M.K., (1934) *Harijan*, 16-11-1934.
9. Gandhi, M.K., (1940) *Harijan*, 27-1-1940.
10. Gandhi, M.K., (1935) *Young India*, p. 321.
11. Gandhi, M.K., (1936) *CWMG*, Vol. 70, p. 295.
12. Kumarappa, J.C., (1945)
13. Kumarappa, J.C., (1936) *Why the Village Movement,* Wardha: Maganvadi, p. 99.
14. *Ibid.*, p. 101.

Preface

The world is a beautiful place to live. With endless pursuit of science and technology, there are still pockets of society untouched by the so-called development process or should we say neglected and tormented by it. Millions of people sleep on the footpaths, and remain hungry nights after nights. Many more have no access to the basic amenities of life: food, clothing, shelter, education, health, security and self-esteem.

The time has come for the development process to go beyond GDP and focus on equity and happiness too. It must take care of those who have been marginalized. It is the law of nature that matter moves to a state of equilibrium. Human society too must move towards equilibrium by bridging the widening gap between rich and poor, educated and illiterate, healthy and sick, and rulers and ruled.

Why should generations suffer because they were born in a particular country, community and family? It is a difficult question to answer unless we are prepared to think outside the ruling development paradigm. The world as it is shaped today has rich nations and poor nations, rich regions and poor regions, rich communities and poor communities and rich people and poor people. Modern technology is often blamed for these imbalances. This book tries to prove that technology *per se* is not the cause of the prevailing mal-development; it is our mindset that is the cause. If we want to use technology for the welfare of all (Sarvodaya, to use Gandhi's term) it would lead to an egalitarian society. But if we use it to serve the interests of a select few, it would produce what we have today.

The book looks at the application of modern science and technology and management techniques to help poorer sections of society earn their livelihood and stand on their own feet. It proves beyond doubt that there is need to balance economic growth with social regeneration by involving the poor in the larger interests of the society; and to rearticulate technology to suit the aptitude and skills of the people. The case studies cover a variety of fields and activities: business, manufacturing, education, women's empowerment, shelter, cooperation, water management, etc. Technology plays a vital role in all these areas. Case studies of people's initiatives in all these areas demonstrate that the poor can contribute as much to the societal development, as the rich, if not more. It points to the kinds of interventions required.

The book is a by-product of the post-doctoral work of the senior author on sustainable technologies carried out at the Institute of Advanced Studies, United Nations University, Tokyo. The study made her realize how business schools the world over were blinded by the narrow vision of profitability, and overlooked the potentials of people's participation and appropriate use of technology in improving the quality of life of those who have been marginalized by the contemporary development style. The scope of the post-doctoral work was enlarged to cover other cases where technology was brought down to human level and make lives of millions more meaningful.

Only those cases have been included in this volume which have been successfully demonstrated in India and can be replicated elsewhere with appropriate contextual changes. It aims at promoting entrepreneurship with a cause and long lasting success, and demonstrates that Gandhian vision of decentralized economy and people's control over technology is not an utopia; it is paradigm which can be used to create an alternate future, a future that is sustainable, humane, and egalitarian.

The book offers insights on the nature of interventions needed, and target groups to be addressed. Contemporary development process has worked through various sectors as if they stand independent of each other. The book suggests Gandhian approach, an approach that is holistic and devoid of ostentatious and wasteful life style. It takes ITC as the means and leads to sustainable development wherein people's participation is assured, ecological concerns are taken care of and grinding poverty of the people is ameliorated. This model can be fruitfully utilized by the corporate and business houses.

Instead of borrowing models from the west where development contexts and conditions are different, India needs indigenous models suited to Indian needs, conditions, and contexts. We need to tap our immense human resource base and initiate development from below involving the people at the grassroots level, the people who are outside the ongoing national development programmes. The book, through a series of case studies demonstrates how such indigenous models can be derived and put into use for the benefit of the poor.

We are thankful to the IIMT for institutional support. But for opportunity Prof. R.P. Misra, Director, International Centre for Gandhian Studies and Peace Research, Rajghat, New Delhi, and former VC, Allahabad University gave us to be part of a larger work he has undertaken under the title Rediscovering Gandhi, this book would have never come out. Special thanks to Robert Bouzon ex-Director University Relations Program at Hewlett Packard for being the mentor of the senior author. We are also grateful to Dr. R. Kapur and Mr. Amit Kapur for their support and encouragement. Mr. K.B. Kachru has always inspired us to do something new. We are grateful to him. We are so grateful to Dr. Kamlesh Misra, former Director, IIMT and Prof. U. Venkatesh former Dean, IIMT for helping us take up and complete this work. Manjit and Amit Saxena have helped us in designing and editing the book.

Gurgaon, NCR
March 5, 2013

Vinnie Jauhari
Parul Wasan

Contents

1

Spiritualising Science and Humanising Technology

"Where the mind is without fear and the head is held high
Where knowledge is free
Where the world has not been broken up into fragments by narrow domestic walls
Where words come out from the depth of truth
Where tireless striving stretches its arms towards perfection
Where the clear stream of reason has not lost its way into dreary desert sand of dead habit
Where the mind is led forward by Thee into ever widening thought and action
Into that heaven of freedom, my Father, let my country awake."

—Rabindranath Tagore, 1913

As I begin to write this chapter, I am reminded of this beautiful poem by Tagore. There are beautiful thoughts of a world full of serenity, respect and peace with dignity for everyone. The great progress that science has made in the areas of medicine, communications, and transportation has led to the reduction of distance and time. But it has not been without a price. The price has been the widening chasm between the haves and the have-nots, emergence of power centres and a violent world where the threshold is very low. It is ironical that with so many accomplishments, the element of rationality has taken a back seat and we continue to be governed by the tyranny of our own making. The ideology of Gandhi during these turbulent times gives some insights unto the role that science and technology can play. Science, on its own, is like a river which flows and emanates energy. It is for human intellect to harness it and for human values to channelise it in the right direction.

All men are born equal. Every human being has a right to dignified life. Strangely, the barriers created by human beings by dividing the world into countries, geographical clusters, colour and caste result in individuals leading happy or horrendous lives. Why should one be punished for being born in a developed or developing country to rich or poor parents? By creating artificial barriers, one cannot prosper at the cost of others for too long. For any society to live in peace, it is essential that there be equitable distribution of resources and interventions are so adapted that the benefits percolate to the people living at the grassroots level. It is important that the basics of life such as food, shelter, clothing, education, means of livelihood, healthcare and a dignified life is available to all citizens. Science and technology have a role in facilitating these accomplishments. However, in today's scenario one sees that there are pockets of poverty in developed countries and even in emerging economies like India and China; there is a huge population still struggling for basics in life. Ironically, India has the world's largest pool of scientific manpower as also the largest poor of the poor. Well, somewhere the benefits have not trickled and this also raises question marks on the agenda of our science and technology mission. It is important to emphasize that there is a need for spiritualised science and humanized technology. The underlying sentiment for this phrase is that science must have a noble intent – of rising above petty interest and working towards the welfare of the humanity as a whole. Humanised technology connotes the sentiments of technology sophistication which engulfs everyone in its fold. It is sophisticated enough to launch satellites and probes to the most distant planet in the universe but humane enough to make the basics of life affordable for every living being.

Technology has a huge potential of touching the lives of individuals as never before. The developments in the area of information technology are fascinating. With the great accomplishments which has led to automated gadgets and VSAT communication, it has also led to certain societal changes which has changed the way people look at things. The investments in information technology by the government has led to an insight into what the government has done for its citizens. There is a higher degree of transparency and, therefore, higher accountability as well. People can view the budget and expenditures and, therefore, have the power to elect more progressive governments. The technology intervention adopted by some of the entrepreneurs in India have resulted in some interesting developments. The e-chaupals initiative by ITC is an amazing story of sustainable entrepreneurship. It is an example of sustainable entrepreneurship at its best.

The whole idea is to use the information technology to generate sustainable livelihoods. The model is simple. They used the existing networks in villages and re-looked at the procurement processes. They identified local farmers who would buy the crop from the local farmers at the best going rate. The farmers would not have to run to various places as the best rates would be available to him because of the satellite connectivity. This would save him time and also give him good return for the labour that he invested on the soya crop. The local farmer would collect the crop from these tiny farmers and then would sell it to *samyojak* from where ITC would buy the soya crop. The advantage of this system is that ITC re-defined the roles of these farmers. It made investments in technology. It employed satellite technology rather than technology which used the conventional cables so that access was faster. As a result of this the farmers started getting better money for their crop. Also they would save their time spent in roaming from one market to another. For ITC, the procurement cost got reduced from Rs. 900 per tonne to about Rs. 450 per tonne. In this kind of an intervention, the roles of the various people were redefined. Technology was effectively utilized to provide better income and in turn the organization also saved lot of money in procurement and used the network created by the use of information technology so that products could be routed through the similar route as a result of which it would make money on any product sold through that route. The investments would be recovered in a span of two years as all major costs are fixed costs. In this scenario, it is a win-win situation for everyone.

Science itself is an unbridled bundle of energy. If one looks at the process of discovery and innovation, one finds that even the person discovering it does not know what the discovery is going to be and the form in which it would be manifested. Also the challenge lies in commercializing it and letting it seep in the market. It is essential that the benefits trickle into the market. So then some of these questions arise—Who are we innovating for? What kind of technologies are required at the grassroots level? What kind of impact can the hi-tech technologies have on the society at large? If one reads about Gandhian philosophy, it gives insight into a vision that transcends time and has lots of implications for human dignity, empowerment of people, equality, self-reliance, care for nature and sustainable livelihood. In his words :

> "It is beneath human dignity to lose one's individuality and become a mere cog in the machine. I want every individual to become a full blooded, full developed member of the society."

The policies which favour the have's against have-not's, are all meant for exploitation. Anything, which leads to exploitation, is not a fair at all. Even when discussing about rural industries, his ideology was that of a village being self-contained. He was not against the use of modern technology for enhancing efficiency. These principles find a reflection in all the community development programmes he initiated. Only those programmes which have community involvement, which lead to individuals themselves as a part of the whole process are genuine.

In this kind of scenario, then the policy of science and technology and its linkages with the larger interests of the society, therefore, have to be kept in mind. It is, important to have a holistic perception of life. Bringing an insight from *Gita* and the Hindu tradition, wherein life is seen as eternal and cycle of birth and death as incidental, one needs to have a sense of concern for continuity and an element of serenity in the surroundings. The basic law of physics is that everything moves towards a state of equilibrium. The unequal distribution of resources, exploitation of one nation by another, exploitation on the basis of colour, caste and religion are all aspects, which have to be dealt with sagacity. The environmental damage caused by some will have far reaching consequences on generations to come. The greed for power and wealth corrupts nations and individuals.

With this as the backdrop, it is essential to look at the role and goal of technology and science in the years to come? Is it a force that can be trapped and channelised? Can we really control it? How can it be used as a tool for creating more equal nations and societies? How can it reverse the environmental damage? How can it ensure a better life for millions steeped in poverty? How can it ensure education, shelter, food and water for everyone? How can it create an egalitarian world?

It is with this intent that this book has been written. The book resonates the Gandhian philosophy and tries to look at some experiments in India which indicate how science and technology has been able to create a difference in ensuring a more equal world and how a sense of empowerment and a more equal society can be created? It is not a utopian target but something which is achievable. It just requires a realization, awakening, and collective action to ensure that these goals are achievable. It is a humble attempt to remind ourselves that we are infinite reservoirs of energy. We all have to collectively assume responsibility for right actions and put in dedicated work for achieving these goals. Nothing then is out of reach. Peace within and peace around is a state which is a goal that one wants to accomplish. The means are also as much important as the goal. So using our science and technology resources with a vision behind can help us leapfrog into a world which is more liveable, more happy and more equal.

Spiritualised science and humanised technology conjure up images in the mind of a pure and ideal state which is possible to achieve. This can be achieved when various stakeholders in a society take responsibility for their actions. Each person or an organization can either take a short term view or a long term view on different aspects. Firms which take a short term view tend to have an *overarching* concern with profits thereby making compromises on processes and ill effects on society. Prosperity at the cost of exploitation can never be long lasting. There are many instances in Indian context wherein there have been market leaders in select industry segments which no longer exist today. They perished by not using adequate systems and consumer promises. The imperialist and oppressive governments have given way to democratic regimes. The basic law of nature cannot be defied in the long run.

A humanised technology implies technology with a human face. It cannot afford to be ruthless and unkind. It brings about development with a concern for basic necessities being fulfilled. The development process can take various routes. The society can be classified in three stratas. There is a strata which does not have access to basic amenities of life. Every human being has a right to live his life with access to food, water, shelter, clothing, health, and self education to ensure that people have access to all of these. There is another strata in the society which has access to basics in life and can afford the luxuries as well. The third strata constitutes the super rich and ruling states. The government needs to ensure good governance and create opportunities to bring the lower strata up and be a part of the development process.

Each stakeholder in the society must assume his role judiciously. There is a need for doing the right things. For instance, let us take the case of a business firm. If it can offer products which are produced by keeping environmental factors in mind and also meets customers' expectations, then the firm would have a lasting life. If one of these is not met, sooner or later there would be chaos. The profitability is important because it helps the firm to stay on and also bring in a pressure to align business processes to consumer needs. The society on the whole can have a great impact on what is being offered to it. There is a saying that nobody can ride on your back unless your back is bent. The people in a society need to understand their rights and responsibilities. Rights cannot be exercised unless one takes responsibilities for ones actions. History is a witness to immense power that society at large can exert. When governments are largely democratic, the power that rests with the society is huge. People need to realise that they have the capacity to bring about a change in the governance of a country. For people to be progressive, they must be educated, healthy, have access to food, water and shelter. There needs to be equality of opportunity for both the sexes.

About the Book

The book is divided into nine chapters:

Chapter 2 on "Stimulating a Ray of Hope at the Grassroots Level" highlights the cases of success pertaining to education, employment, housing, water, health and empowerment of women. The chapter focuses on the case study of the Social Work Research Centre through its Barefoot College programme and targets interventions at an illiterate community thereby generating literacy, employment, water and food. It also focuses on initiatives undertaken by SEWA to bring about women empowerment. It touches upon initiatives in the area of housing by NIRMITHI and SPARC. The chapter highlights the use of elementary technologies to initiate grassroots interventions. It makes use of local knowledge and community involvement as essential to make the intervention successful. It clearly brings out the fact that there rests huge amount of power with people. It just requires the leadership effort of few individuals to awaken the masses. The need for identification, leading by example, willing to live and associate with the community is essential to make the intervention successful. The role of government is also highlighted. There is a myth that even corrupt governments nurture. Wealth generated at the cost of poor or by exploitation cannot be sustained. Progressive governments not only create wealth for the society but also open doors for their own sustainable development.

The chapter 3 on "Sustainable Empowerment of Women" discusses the issues of women empowerment. It traces the gradual decline of status of women in Indian society. Examples of different interventions on women empowerment using varying levels of technology have been given. The case of SEWA has been discussed and the range of activities undertaken by them have been elaborated. This section also develops on the case of Sri Mahila Griha Udyog—Lijjat Papad. It is a case of village based entrepreneurship which spread itself in early 1950s. The interventions range not only to commercial operations promoting a brand called 'Lijjat Papad' but also generates employment, contributes to women literacy. The case study again demonstrates the use of basic technology in bringing about empowerment of people.

The empowerment of women is one of the critical links in the entire development process. Literacy of women help them contribute to overall health and hygiene standards in the family. It has a bearing on children's education and family income as well. It nurtures a spirit of freedom and independence. In this intervention again, the vision of a single individual has played a very important role in the creation of the entire set-up. The intervention has not been very focused to a single sector. The interventions have stretched beyond a single element of women empowerment. It simultaneously includes the role of various sectors in making a difference to the lives of people at the grassroot levels.

Chapter 4 on "Towards Sustainable Livelihoods" discusses the interventions in a rural community tapping the source which is abundantly available. This chapter discusses the initiative involving the entire community. It discusses the "Operation Flood" which ushered white revolution in India.This initiative has been one of the largest dairy development programmes in the world. It has also been responsible for setting up extensive and most modern facilities in milk storage and transport. Here again, this initiative was conceived as a means to generate employment and self-esteem. It helped the community to upgrade their lives and improve their standard of living. By investing in basic technologies, employment was generated. It resulted in a sustainable experiment which was capable of being replicated. It also reinforces the indigenous strength of people and their immense capacity to organise things. Structure and systems can be created and adhered to if people are committed and motivated. It brings out the role of a basic leader and a firm which takes such an initiative. The chapter also discusses the role of education in generating sustainable livelihoods.

Chapter 5 is on "Sustianable Health for All" tells that Gandhiji laid greater emphasis on maintaining health than treating diseases. Health is immensely valuable and care must be taken to ensure that community should be healthy. In India, there is a lot that is still desirable as far as availability of health care facilities is concerned. The public spending patterns in health are not very pro-poor. Even internationally, preventive and promotive health services take a back seat as compared to curative case. This chapter looks at some of the interventions such as Sulabh International. This intervention was initiated to introduce effective public sanitation services. The intervention deployed low cost technology and easy to construct locally available material. It also discusses the case of Nalamdana. Its mission is to provide health education and promote preventive behaviours through entertaining methods: street, theatre, discussion groups and songs with messages. It also discusses another intervention, "CHIRAG" (The Central Himalayan Rural Action Group). The intervention has focused on women in child bearing age. Another intervention discussed is the "Jaipur Foot". Here Indian Space Research Organization transferred its polyurethane technology to a Jaipur based social organization. This technology is a spin off borne out of research and development of various polymeric materials. It helps to produce more durable, lighter and comfortable artificial foot.

Chapter six is on "Corporate Social Responsibility : Some Interventions" examines the role of the corporate firms in humanising technology and also ensuring their long term survival. Firms internationally are realising the importance of corporate social responsibility and its link with profitability of the firm. Gandhiji's ideology has lot of substance wherein he felt that mankind was the custodian of the natural wealth and, therefore, had no moral right to destroy it or use it selfishly. This chapter discusses the interventions initiated by M.S. Swaminathan Foundation. The MSSRF has been consciously designed as a scientific and training institution "without walls". It is an institution which relies on partnerships for its vitality and effectiveness.

Chapter seven is on "Role of Corporate Sector in Sustainable

Development: Lessons from Success Stories", takes the previous chapter forward. It blends the elements of profitability with sustainable development. It tries to link new product development by meeting the needs of the people at the grassroot levels by utilising local resources. These interventions are either initiated by the business firms or NGOs. The technologies used range from being sophisticated to a very basic. However, the unique aspect is that these technologies can be very easily utilised by the people at the grassroots levels. The interventions that have been discussed are the ITC e-choupal initiative. It is a case of a tobacco major and a diversified firm exploring possibilities of conducting business with a rural community. This is a vital link in supply chain management. The intervention has resulted in major cost savings in the procurement process. The volume of new business is so large that even if the tobacco business were to stop, it would not have any impact on the overall business of the firm. It is an intervention taken up by a firm by redefining the role of the community members. It is astonishing to see how quickly people adapt to new and sophisticated technologies if it improves their livelihoods. The chapter also discusses another case, "Tarahaat". It offers various services such as Tara Dhaba, Tara Bazaar, Taradak and so on. The intervention has a wide application ranging from farmers, traders, housewives, senior citizens and children. One of the corporate Indian firms — Tata's corporate interventions have also been discussed. The chapter also discusses Hewlett Packard's Global Citizenship Initiative. It links the product development with job opportunities for women and men in rural communities. It creates opportunity for employment for people and also generates profits for itself.

This chapter then looks at various interventions and makes a comparative assessment with respect to:

1. Objectives;
2. Stage of development of the community where it was targeted;
3. Geographical reach of the intervention;
4. Nature of beneficiaries;
5. Financial feasibility of the intervention;
6. Dependence on grants;
7. Ability for self-sustenance;
8. Range of intervention varying from being specific to being very wide in its domain of operation;

9. Domain of intervention; and
10. Networking with other stakeholders.

The eighth chapter is on "Towards a New Framework for Sustainable Development". This chapter links the nature of technology used with the nature of development of geographical area where the intervention is targeted. The technology varies from being low tech, medium tech to being very sophisticated. Interventions such as Barefoot College can be initiated with low levels of technology. This can graduate to a better framework deploying medium level of technology. Experiments such as those initiated by Hewlett-Packard P and Swaminathan Foundation fall in this particular category. Experiments such as those initiated by ITC are characterised by deploying high level of technology sophistication. These interventions need to be self sustaining in order to survive. They have to move out of the purview of a single entrepreneur and need to be institutionalised. The experiments have to involve community for successful implementation. Such interventions are also characterised by network effects. These interventions require a very close networking of the community, NGOs, government and the business firms.

Chapter nine, sums up the Gandhian philosophy and prescribes the action plan for various stakeholders – government, civil society, business firms, NGOs. Recommendations have been made in relation to women, sustainable livelihoods, education among other aspects. The chapter also delineates on how spirituality can contribute to direction in our work and personal lives. The book suggests that a selfless spirit with a determination for hard work is an underlying value which all stakeholders will have to adopt.

This book is a journey of technology which cares and nurtures. It is a story about the power of people and the ability to work in teams. It is a saga of how people can live with self esteem and pride. It also exemplifies, how profitability by business firms is possible, when they link their products and processes with people at the grassroots level. The need is to widen the number of such interventions and make them sustainable. These efforts need to continue and fundamental rights of individuals need to be respected and upheld. The basic right to live a life with dignity and self respect is essential for all citizens. The chapters unfold the story of technology with a human face.

2

Stimulating A Ray of Hope at the Grassroots Levels

For technology to be meaningful, it is essential that there is a ray of hope for people at the grassroot levels. Gandhi had conceived of an egalitarian society wherein there is self-reliance, dignity for work, no discrimination based on gender or colour, having a vision and being self-independent, utilizing existing strengths and resources. These are thoughts that have a timeless appeal and are as much needed as they were needed centuries back. The mindless pursuit of power for sustaining and nurturing personal ambition tilts an individual and, therefore, a society towards exploitation and greed. Ultimately any amount of power and accomplishment must result in a situation which yields happiness and peace. Technology in today's society has led to disequilibrium on power structures. This has fundamentally led to the emergence of developed and developing countries. The huge digital divide would make the situation worse. The need of the hour is to invest and channelise technologies for empowerment of people, initiate interventions so that these could affect the lives of people positively. This chapter illustrates some of these interventions initiated by numerous organizations on interventions happening in various areas. These interventions have some pointers for the technologies that are being used and how they impact the lives of the people.

The following paragraphs highlight the interventions taken at the grassroot level. These cases pertain to the education, provision of employment, housing, water, health and empowerment of people, also reflecting specific cases on empowerment of women. The Social Work Research Centre (SWRC) through its Barefoot College is an example of how a local community which is completely illiterate, could be used for employment, can be educated, could use local knowledge and technology for water and food, can manage itself as a sustainable unit. Similarly, SEWA is another case which empowers the poor self-employed people and the interventions cut across different sectors such as shelter, employment, health, education, crèches, standing up for the cause of self employed workers and raising issues on their behalf. The case study on SPARC focuses on the use of local technology in the area of shelter and empowerment of women. The Building Centre Movement also focuses on the provision of the shelter to the poor communities. The case of Sambhav highlights the use of local seeds and indigenous knowledge in converting a barren land for creating sustainable livelihoods. The Barefoot College exemplifies how education has to be rooted into the ground reality of working children who have to support their parents for livelihood. It runs night schools and also empowers children by giving them responsibility for various tasks.

SWRC INTERVENTIONS IN PROVISION OF BASIC NEEDS

The Social Work Research Centre (SWRC) has been working in rural communities in India to improve the quality of life of the rural poor. SWRC has worked to address basic needs: water, health, education, employment, social awareness and conservation of ecological system, by enrolling individuals in the processes that govern their lives. SWRC is a voluntary agency with its main centre is in the village of Tilonia in Silora Block in the Ajmer district of Rajasthan. The organization began its work in 1972 in Tilonia, by opening a Barefoot College. Rural development required living among those people who would effect and be affected by that development process. SWRC programmes were initially started with urban expertise from outside the area.

The College benefits the poorest of the poor who have no alternatives. It encourages practical knowledge and skills rather than paper qualifications through learning by doing process of education. The College is spread over a 60,000 square feet consisting of residences, a library, dining room, meeting halls, marketing outlets, an open theatre, a blacksmiths' workshop, solar fabrication workshop, water testing laboratory, an audio-visual unit, handicraft production centre, a puppet workshop and a 4,00,000 litre rain water harvesting tank entirely built and supervised by the local people. The College serves a population of over 1,00,000.

Philosophy

The philosophy at the Barefoot College is that people are encouraged to make mistakes so that they can learn humility, curiosity, the courage to take risks, to innovate, to improvise and to constantly EXPERIMENT. It is a place where all are treated equal and there is no hierarchy. The idea of Barefoot College is based on the belie that development programmes do not need urban-based professionals because para professionals already exist in the villages whose wisdom, knowledge and skills are neither identified, mobilized nor applied just because they do not have an educational qualification.

The Barefoot College is developed in the following manner :

1. First, by reposing faith in the competencies of rural poor who have the knowledge, the skills, the wisdom and the desire to identify and solve their own problems.
2. Second, by creating an environment and a situation where these skills and knowledge can be applied for the community's own development.
3. Third, by informal, non-structured, on the job practical training until such time as the persons have acquired the confidence, the competence and the capacity to provide the service without any help from outside.
4. Fourth, by choosing an area which is remote, inaccessible and very difficult physically to reach, so that there is peace, mental space and non-interference from the so-called experts who are dying to make sure you fail.

People who have no formal educational or professional degree run today all the programmes. An individual's will to learn and aptitude for learning is more important than any formal degree or paper qualification. The new campus at Tilonia was designed and built by one of the villagers, who could barely sign his own name. The campus itself reflects the adaptation of both traditional as well as new methods and technologies. Old traditional methods have been used to keep the buildings cool while solar energy is used to provide electrical power to the campus. People with minimum paper qualifications work as night school teachers, health workers, computer operators, solar engineers or hand pump mechanics. Basic literacy, health and first aid skills are also taught. In this way each individual learns about the entire organization, its mission and its functioning.

Sustainability

1. It generates employment,
2. It involves people in the process, and
2. It helps to take care of the basic minimum needs.

The Centre does not provide free services. A nominal fee is charged for all services including health services, training, and installation of hand pumps or solar electrification for lighting. Almost 98 per cent of the workers are from neighbouring rural areas.

Technology Orientation

SWRC believes that new trends in technology and high tech machines are not always synonymous with development. SWRC does not believe in imposing technology on people in rural villages or using technology, which deprives people of employment. Adapting and improving on pre-existing, traditional ways is often more effective than using newer technologies.

Replicability of the Barefoot Concept

1. It will work anywhere in any poor rural community anywhere in the world where there is extreme poverty.
2. The rural communities are neglected, deprived and forgotten so they have no choice but to develop and depend on each other and not on people from outside—thus all knowledge and all skills are useful, necessary and respected.
3. Where the percentage of ILLITERACY is high so the oral tradition is rich and knowledge skills are traditionally passed down from one generation to another.

The Barefoot College encourages the following people to participate :

1. Those who are dropouts, cop-outs, washouts and who are rejected by the society because they cannot pass the exam and have a degree.
2. Those who have no possibility of getting the lowest of the low government job. They have no choice but to stay and

the investment in the training is not wasted. They will earn the respect of the communities they serve because of the service they will provide.

Organization

The Barefoot College concept is percolated to the communities in the 110 villages of the Silora Block through the 12 SWRC field centres. The field centres have the freedom to decide their own course of action. Each serves between 9 and 35 villages. The College has over 400 staff members working full time in various activities related to basic services. They have no formal qualifications for the job they are doing. With the help of a cadre of barefoot engineers, doctors, teachers, designers, chemists, accountants and traditional communicators, communities are using expertise they acquired from their ancestors. The concept of communities depending on themselves has revived. Indigenous institutions and decision-making processes have been activated and villagers have gained new confidence. They increasingly recognize their own strengths and assign value to their own skills—something that was never felt before.

The Barefoot College and Children's Parliament of Tilonia, Rajasthan has won the Children's World Award (Sharma, 2001). The award, considered the "Children's Nobel Prize" is a unique global award for organizations that champion the rights of children through their activities. Queen Silvia of Sweden at Grips Holms Castle in Marie Fred presented the award on April 18, 2001. The award carries a prize of $12,500. The prize money is to be spent on activities conducted by the College for the rights of the children.

Uniqueness and Importance of Barefoot Education Initiative

The efforts of the Barefoot College are commendable as the education infrastructure is in a deplorable state. The quality of school infrastructure leaves a lot to be desired. The fifth All India Educational Survey conducted in 1986 found that 40 per cent of schools did not have a permanent building and 9 per cent did not have any building at all (NCERT, 1992). As far as the availability of other basic facilities was concerned.

60 per cent of schools had no drinking water, 89 per cent did not have toilets, 40 per cent had no blackboards and 70 per cent

had no library. Six years later, there is little change (NCERT, 1997). As government and local bodies run the majority of the schools (92%), the state has primary responsibility for bettering the quantity and quality of schools. The problems of physical distance for children from a school are exacerbated by problems of social distance for children from underprivileged households (Probe Team, 1999).

Swaminathan & Rawal (1999) highlight the case of an experiment taken up in Madhya Pradesh where an Education Guarantee Scheme was introduced in 1997. They quote Gopalakrishnan & Sharma (1998) and mention that the government guaranteed the provision of a teacher, teacher training and basic materials within 90 days of a request from a community for a primary school as long as there were a minimum of 40 children (25 in tribal areas). It led to setting up of 15,568 schools in one year. On the negative side, the small expenditure on the scheme (teachers monthly salary of $10 a month and children given free books worth half a dollar). Swaminathan & Rawal (2000) point out two major reasons for failure of this scheme. There was no commitment of additional financial resources for the scheme. Second, it required a local community to start a school and run it successfully for two years, with some minimal help from the government in terms of teaching materials and other unspecified assistance before the government upgraded it on a permanent basis. It is not clear if other assistance includes even basic salary for teachers and some teacher training.

Nearly 38,000 additional posts of teachers'are lying vacant in primary and upper primary schools in various States with hundreds of crores of rupees remaining unutilized (*Hindustan Times*, 2000). Against 83,045 sanctioned posts for third teacher in primary schools, the States have so far appointed only 74,463 teachers. Till March 10,1999 as much as Rs. 3,78,69,000 out of Rs. 25,90,16,00,000 under Operation Black Board scheme were lying unutilized. The Parliamentary Committee says that this situation calls for a serious review. Even in 2005 the situation was very different.

In the light of the above mentioned facts, the interventions initiated by Barefoot College hold a lot of merit as they involve local community, are flexible keeping in mind their commitments to generate food for themselves. It also gives them practical knowledge, which helps them use it on the job.

Interventions in Education

The College runs a series of night schools in several villages in and around Tilonia where children are taught in the evenings, after they finish their day's work. The college students have their own parliament, the representatives to which are elected from amongst the boys and girls attending the different night schools.

Night Schools

More than 80 night schools have been set up for the benefit of working children. Nearly 1200 girls and over 1500 boys who tend cattle during the day attend these schools after dark. Solar lanterns maintained by rural solar engineers power more than 68 per cent of these institutions. All teaching aids and learning materials used in the night schools are made from waste materials. Instruction is informal and curriculum is focused on practical knowledge and experience. Since most children tend cattle, they learn basic husbandry along with maths. they attend night schools for five years. They monitor their own schools by electing their own representatives.

Children's Parliament

The children's parliament controls and supervises the night schools. It is based on the belief that giving power to the people who have a vested interest in the school is the best way of ensuring its success—as well as making the children aware of political structure and processes. This form of education related activism provides a heightened awareness of the system, its workings and avenues for redressal of local grievances.

There is also a provision for teacher training and there is a mobile library which goes from village to village and from where the children of the night schools can borrow books.

Concrete Outcomes

Hand Pumps

More than 1500 India Mark II hand pumps were installed between 1979 and 1995. Over 3,00,000 people continue to benefit from these hand pumps. Despite claims by the government that it was technically impossible, 28 hand pumps were installed at 15,000 ft above sea level in Ladakh and operate at –40 degree centigrade.

Since 1993, the College has focussed on water harvesting and piped water systems as the emphasis has moved beyond clean water to easy access to drinking water. 12 villages, 12,000 connections and 15,000 people now benefit from community piped water supply, designed, planned and implemented entirely by the village people. These communities pay Rs. 20 per month for two hours of water per day (http://www.barefootcollege.org/html/water.htm.).

Other water initiatives include :

1. 12 million litres of rainwater were collected in 155 schools and community centres. This water is the only safe option in areas of brackish water with high iron and fluoride content,
2. 1400 samples of drinking water covering 78 villages in 8 States were tested using mobile testing kits,
3. 35 rural youth were trained as barefoot chemists in 8 States,
4. 753 hand pump mechanics were trained to carry out all repairs for the 45,000 hand pumps in Rajasthan, 40 of these mechanics being women.

Contribution to Health

Health

Barefoot College has remarkable achievements in the health sector too. Begun in 1973, the health centre served as a small dispensary. Villagers are now charged a nominal amount for medicines. A team of doctors pays regular visits to villages for routine health examinations. Today more than 200 health centres serve villages throughout India. Since 1986, the Barefoot College has been using biochemic medicines. Biochemic medicines are a set of 12 medicines, which can be combined and used for different ailments. The college has developed 28 medicines using the twelve core medicines. At least one field staff member in each village has been trained in this alternate system of medicine and serves as the field centre's *svasthya karyakarta* or health worker.

Today, more than 200 health workers serve a network of Indian villages trained to tackle the health issues and minor injuries. The health workers can give artificial respiration in emergencies and take a patient to the nearest government hospital when necessary. Health workers also teach villagers about basic health issues including hygiene, the importance of vaccinations and other preventive measures. Barefoot midwives are trained in proper delivery methods as well as pre- and post-natal care of the mother and the child. In case of birth complications that cannot be tackled in the village, the *dai* escorts the expectant mother to the nearest hospital.

Environmental Initiatives

Barefoot solar engineers have installed solar photovoltaic units across 10 States of India in 300 adult education centres. The results include:

1. 500 solar lanterns manufactured at the college for 200 night schools across the country,
2. 104 fixed solar units for night schools to replace kerosene lamps, which have a negative impact on children's eyesight,
3. 25 remote and inaccessible villages in Ladakh have 36 kws of solar panels that provide three hours of light in the bleakest winter to 930 families,
4. In Leh and Kargil districts, solar energy initiatives have saved a total of 59,000 litres of kerosene,
5. 79 rural youth we have been trained as barefoot solar mechanics with absolutely no aid from urban professionals,
6. 1,30,000 litres of kerosene saved, by replacing generators and oil lanterns with solar power,

Funded by the European Commission, the College is working with Programme Asvin to develop and disseminate solar energy systems for villages in the Himalayan region of India. The project is bringing solar powered lighting to 30 villages in Sikkim, Uttar Pradesh and Ladakh—as well as demonstrating how local knowledge and practical skills can make these villages completely self-sufficient technically and financially.

Wasteland Development

Most of the land owned by the government or village and reserved as grazing is wasteland. In wasteland Rajasthan is barren because of overgrazing and desert like conditions. The Barefoot College helps rural communities to regenerate this land. The College provides seedlings from its nursery of draught resistant trees, shrubs and grasses. The villagers themselves plant the trees and shrubs, which become a source of fuel and fodder. Every wasteland has a watchman who prevents trespassing or misuse.

The Barefoot College aims to drought-proof these areas by employing various strategies:

1. Wasteland development through plantation,
2. Popularizing traditional systems of water storage,
3. Recharging old wells from rain harvesting units, and
4. Preserving desert culture and mobilizing people's actions.

The rain harvesting initiatives have achieved :

1. 207 underground tanks with a total capacity of 11.5 lakh litres built for rain water collection in Rajasthan thereby employing 4000 persons,
2. 12 million litres of water collected in 1996-97 in rural schools and centres where rain water harvesting units have been installed,
3. Because of the availability of potable water, attendance of girls in these schools have increased significantly,
4. 2,325 landless labourers have been given 93,500 days of employment to build these tanks, and
5. 90 lakh litres of water was collected at Re. 0.25 per litre in 1996-97 with individuals contributing over 5 lakhs worth of their labour.

Water Interventions

Lack of access to clean water and sanitation is a problem that affects large number of people. Water is a wholly renewable resource. This may be true in a geological time scale but in a yearly or even centennial cycle, this is not necessarily the case. Most surface water does replenish itself on a weekly to yearly basis but ground water, which is a major source of fresh water, has an average renewal cycle of 1,400 years (UNEP/GEMS, 1991). Without artificial aquifer to recharge, ground water reserves cannot be considered completely renewable and the use of water from them reduces the overall supply. The increase in demand for freshwater has resulted from a rapid growth in urbanization and industrialization combined with an overall increase in population, both rural and urban (Martin & Martin, 1991). In many countries agriculture can account for upto two-thirds of human water demand. India is reported to use 97 per cent of its fresh water supplies for irrigation (UNEP/GEMS, 1991). With 21 per cent of the world's population, China has to survive on only 7 per cent of the world's total fresh water resources. Some 300 major Chinese cities face water shortages (Glenn & Gordon, 2000). Japan changes seawater into drinking water and exports that technology to the Middle East (Glenn & Gordon, 2000).

Water is a public good, which if maintained unspoiled is a vital free good for all human beings. Well known in economic literature, the main cause of pollution in general is the fact that the industrial processes are not based on full cost pricing (Henderson, 1999) including environmental and social costs of the industrial activity in the production of private goods. There is no doubt that if we can create new institutional designs that can bring the right incentives for industry to develop technologies that conserve this vital public good, technology may have a huge impact on the well-being of the poor.

The technological advances need to be channelised in a manner that ensures safe drinking water for the masses. Water is a natural product. It cannot be a monopoly of rich alone. As individuals do we have no obligation to ensure that everyone gets his share of clean water and clean air?

About 70 per cent of the world's fresh water supply goes to agriculture, a figure that approaches 90 per cent in Asian countries such as China and India, which rely on extensive irrigation (Shah & Strong, 2000). The shortage of fresh water is looming large as a serious obstacle to food security, poverty reduction and protection of environment. Eventually, the shortfall could reduce India's harvests by as much as 25 per cent. Reduction in morbidity from better water supply and sanitation is estimated to be 26 per cent for diarrhoea, 27 per cent for trachoma, 29 per cent for cariasis, 77 per cent for schistosomiasis and 78 per cent for dracunculiasis. Mean reduction in diarrhoea-specific mortality can be 65 per cent, while overall child mortality can be reduced by 55 per cent.

Case of Rajasthan for Water Intervention

According to Prayatna Sansthan more than 30,000 hand pumps are being maintained by over 1,000 Hand Pump Mechanics (HPMs) who have replaced government caretakers, block mechanics and the mobile maintenance unit. This decision to scrap the top heavy, prohibitively expensive, three-tier system (designed originally by UNICEF) in favour of an entirely community based repair and maintenance system, was taken years ago by the State Government of Rajasthan. The idea of barefoot mechanics—who have been given adequate training in repairing and maintaining hand pumps in their own villages—appealed immensely to the Government of India. It is the only system recommended by the Prime Minister's Technology Mission. The initiative for the HPM scheme came from the villagers themselves. The question was what was so special about the hand pumps that they require an entire government department and a battery of personnel and equipment to repair them. So the use of the Training of Rural Youth for Self-Employment (TRYSEM) was made to train. The HPMs chosen under the TRYSEM programme were mostly semi-literate and landless youth, the majority being agricultural labourers from the poorest families in the villages. What they possess are practical skills picked up from working and improvising in the village. Unlike the engineer, who does not really suffer if the handpump is out of order, the HPM has a stake in the efficient working of this scheme. He is answerable to the community and earns respect by providing a vital service. He is not a government servant but has his roots in the village and has no choices but to stay.

As a contrast to this in the government run three-tier system, the caretaker provides a free service and depends on the Block mechanic who in turn depends upon the mobile maintenance unit. The fact that the caretaker has only a limited function while the others draw large salaries reflects the fact that there is a total ignorance of the availability of the skills in the rural area.

Another case of water management gives insight into the management of the local technologies. In Dadu Block (in Ajmer district) around the Sambhar Lake, the largest salt-water lake in Asia, the water is at a depth of more than 40-50 feet, is saline and, therefore, unfit for human consumption and irrigation purpose. The entire land is barren but the forest cover is very marginal. The only source of water are open ponds and shallow wells that get filled during the rainy season between July and September. The average rainfall is merely 200 to 400 mm per year. By April the ponds are virtually dry and women have to walk 4 km. to fetch water. Animals and human beings are forced to drink water from the source thereby leading to cause of infections.

Collection of rainwater off the roofs, to be stored and used for drinking water purposes seemed the only solution to the drinking water problem in this harsh terrain. This was an option, which could be easily managed and controlled by the village communities themselves. So Prayatna Sansthan began constructing roof water harvesting tanks in 1986. They were first built in schools in order to provide the children with safe drinking water. The attendance in these schools better because there was now clean water available freely. The tanks constructed as of 1992 have an average capacity of 30,000 litres and are used not only by the children but by the rest of the village community too.

The run off water from the roofs drains *via* pipes, which in turn, are connected to an underground tank. The roof and its walls are repaired and made waterproof and are kept clean. There is a hand pump attached to the top of the tank to prevent contamination by dirty unhygienic buckets being dipped inside. Five hours of heavy rain on 600 square feet of catchment area is adequate to fill a 40,000 litre tank. The experience of the schools has been that this volume provides enough drinking water for 50 children over 12 months.

The tanks are constructed using limestone, bricks, stones and cement. The cost of construction is under Re. 1 per litre. Skilled masons, to be found in almost every village, are the only experts who are required. Over a period of three years, the workers of Prayatna Sansthan have overseen the construction of over 30 tanks in as many village schools. They harvest 15,00,000 litres of rain water in a year.

Organic Farming in Orissa

This case illustrates how communities can be self-sufficient on the food front. The organic farming, soil and water conservation models demonstrated by Sambhav are remarkably low cost with the use of simplest of techniques. The farming can be done with the help of materials that are freely and abundantly available to any farmer anywhere. In the 90 acre Sambhav farm near Odogaon in Orissa, economist and environmentalist, Prof. Radha Mohan, has shown to the world that it is possible to make degraded soil yield gold without the help of irrigation, chemical fertilizers or pesticides or machines and with little resources that are available to small and marginal farmers. The soil of Odogaon was most eroded and highly degraded. He took up challenge because if the experiment would succeed here, it could succeed anywhere. To begin with huge quantity of weeds like Sabai grass in the area for soil and water conservation. The vegetative bunds gave a healing touch and the shallow channels, trays and pits dug around the plantation area ensured that the trees absorbed each raindrop. Percolation tanks ensured that rain or no rain, the soil would not go thirsty. Also there are a series of percolation tanks. So whatever water is collected there, it seeps through the ground providing moisture to the root zones of the plant. So, instead of water flowing on the surface removing the top soil, water flows underground. To keep a tree alive during peak summer, Prof. Radha Mohan would use an earthen pot half buried in soil and with a small hole at its bottom. When water is poured into it, it supplies water to the thirsty roots drop by drop for seven days. Even the weeds that grow can be cut and dried and used as mulch, a source of manure and moisture. Cutting weeds, putting them as mulch, using compost and using green manure – these are within the reach of the poor ordinary people. This technique is capable of being replicable and can be adopted by ordinary people. Because of this experiment, the farmers have stopped selling their lands. The experiment has been sustained without any major funding. The land for the 90-acre farm was bought for only Rs. 90,000 ten years ago. Sambhav has been able to generate revenue by sale of its products and saplings.

Housing : Nirmithi Initiative

Shelter is a basic human need, which needs to be fulfilled. Almost half of Bombay's 8.7 million people live in dirty slums. About 300 families are added to the city's population everyday. Despite the fact that an average of 20,000 to 25,000 housing units are built each year by public and private housing agencies, it is not possible to provide houses for 4 million slum dwellers. The system does not provide the resources required to build for those with limited income and little or no savings at all (Anzorena, 1994).

As shelter is a basic need, cost effective housing technologies become important for low income people. There are problems in the provision of housing for the poor. Important among those are:

1. Rising costs;
2. Poor access to construction materials;
3. Lack of diffusion of innovations to the common man;
4. Awareness these innovations in the professionals;
5. Lack of exposure of innovations among the construction workers and artisans;
6. Adverse impact on environment due to the depletion of natural resources; and
7. Lack of support through building regulatory authorities, codes and schedule of rates.

The Building Centre Movement in India has emerged as a grassroot level intervention with the objectives of :

1. Transfer of technology;
2. Training of artisans;
3. Production of elements; and
4. Construction and guidance.

The Building Centre Movement has taken long strides. From Nirmithi Kendra in Quilon in 1986 in Kerala to Jammu in 1995, there are 385 building centres.

The Government of India launched the National Network of Building Centres. The case of Nirmithi Kendra in India exemplifies how shelter technologies make a difference to the lives of the people who live at the margins. The case is derived from the MOST Clearing House Best Practices.

In 1985, for providing affordable solutions to housing, India's first "Nirmithi Kendra" was set up in the Quilon District of Kerala by the then District Collector. This was a trendsetter in cost effective and environment friendly (CEEF) building technology saving about 30 per cent of the cost. The movement succeeded in technology transfer from R&D institutions, in training and employment generation and in developing new educational programmes. The achievement of this project is that in 1985, it provided shelter to thousands rendered homeless due to calamities by integrating beneficiary participation with appropriate technology and available resources with the district administration.

The factors contributing towards its success are:

1. Beneficiary involvement in the process;
2. Delinking developmental task from government rigidities;
3. Establishing linkages with the R&D institutions for technology selection and transfer;
4. Effective co-ordination of tasks and related agencies;
5. Use of locally available and innovative materials;
6. Cutting down the consumption of energy intensive materials (cement steel) and using appropriate technology;
7. Ensuring local participation in construction activities;
8. Blending new styles with traditional ones; and
9. Designing according to the lay of the land.

The Nirmithi concept soon spread across the State and the Government of India recognized the concept by including it in the Union Budget and the National Housing Policy saw the growth of the movement at the national level by setting up Building Centres in the country. A Special Habitat Award was given to the progenitor of the movement and the Kollam Centre gave further impetus to the spread of the Movement. International recognition was awarded to Nirmithi when the United Nations Commission for Human Settlements at its 14th session in Nairobi (May 1993) adopted a resolution recommending governments to set up institutions modelled on the Building Centres at the national, provincial and grassroot levels. Nirmithi has become synonymous with cost effective environment friendly building technology. To ensure quality criteria, Nirmithi in collaboration with the Bureau of Indian Standards, compiled standards and specifications for cost effective building materials and techniques.

Lack of adequate skills for the use of new technology led to investments in skill upgradation programmes in masonry, carpentry, plumbing, landscaping and such other skills related to housing and habitat. The training activities of Nirmithi include the revival of traditional architecture with its blending with the modern. Young carpenters and craftsmen are trained in dying arts such as hand carving and traditional roofing with stylistic features. A considerable portion of Nirmithi efforts lies in the propagation of CEEF technology by setting up of Nirmithi Clubs in educational institutions. About 2,500 students in 25 colleges are setting up Nirmithi club activities. These clubs aim to generate the right attitude towards cost effectiveness and environment friendliness through the use of poster campaigns, demonstration programmes, seminars and workshops for generating awareness among the students, study tours and field visits, career guidance, entrepreneurial development and various short duration skill development training programmes.

The training centres have contributed considerably to alleviate poverty. Hundreds of youth and women trained in building material production were productively utilized at these centres. Employment generation through these production centres in rural areas has helped to arrest rural migration. For common man needing guidance and information pertaining to CEEF technology, Housing Guidance Centres were set up by way of consultancy, design, estimation and execution as required by the client.

The efforts of the institute have been lauded through tax waivers by the government on Nirmithi buildings and industrial estates. Central government issued orders to waive excise duty for cost effective materials and instructed State governments to execute 20 per cent of public works through Nirmithi Kendras, accreditation and incentives being given to industries using Nirmithi materials, Tamil Nadu State government issuing orders to execute all construction works under the District Administration through Nirmithi Kendras.

Concrete Outcomes

1. 30 per cent reduction in construction cost;
2. From 14 building centres in Kerala to 350 building centres in the country;
3. In Kerala alone, 38.58 man years of onsite employment and 61.74 man years of offsite employment generated;
4. Targeted different groups such as scheduled castes and scheduled tribes and women;
5. Educational programmes such as Masters and Diplomas in Habitat Technology;
6. The Nirmithi Eco Village at Mannanam, Kerala, demonstrates the possibility of living in harmony with nature. Environment friendly houses, non-conventional energy generation using biogas, solar and wind energy, cultivation using organic manures and herbal pesticides are being practiced here. Eco-

friendly income generating programmes give thrust to the idea of sustainable development; and

7. Another demonstration of the Nirmithi activity has been the rehabilitation of an entire community in the coastal area of south Kerala whose main activity was the brewing of illicit liquor, through concerted action including training programmes in building material production and the eventual involvement of these people in Nirmithi production units for actual production – thus empowering a whole community towards positive action and meaningful integration into society.

The Case of SPARC

SPARC is a Mumbai based voluntary organization working with slum and pavement dwellers. The real challenge is in documenting the experiences of women seeking shelter, which can survive delays, long waiting periods and prolonged negotiations. A movement, which ensures that women are in the centre-stage and fully involved in the process that educates and trains people in the process. The model process started in Bombay in 1984, initiated by a group of poor women in Byculla. SPARC was set up to explore ways by which a group of professionals could work along with poor community to resolve problems they felt were critical. It was intrinsic in this aspiration that women would be central to the process. SPARC began to work with women who resided on pavements in central Bombay, seeing them as the most vulnerable group in the city. Along with 600 women residing in five settlements, SPARC explored why poor people could never get secured housing in the city and despite the evidence to the contrary designed a training programme which equipped women to create human and financial resources to make an alternative possible. In this exploratory process an organization called Mahila Milan was formed and a three-way alliance with SPARC, an NGO, and National Slum Dwellers Federation (NSDF), a federation of slum dwellers across India was initiated. It began to provide exposure and training to communities who were members in the federations, also assisting women in these settlements to form Mahila Milan collectives and negotiate space for participation in community matters. The process required constant dialogue with the State officials preparing them to understand the value and advantage of dialogue to the achievement of their work goals rather than as a favour to poor communities. It also meant tremendous participation in the dialogue with confidence. There is a need for a change, which not only emancipates the woman but the entire family. The NGO is a facilitator; it never withdraws; it transforms its relationship with communities and increasing takes responsibility of training and capacity building, and planning projects. Free execution of projects is undertaken by the community leadership, never by SPARC. And, most important, most of these trainers are women.

Impacts

1. Mahila Milan has a standardized shelter training process and trains communities all over the country.
2. Women in small communities are running credit and saving groups, which have convinced banks to lend them money on easy credit.
3. Mahila Milan has undertaken construction in all sites where communities get land tenure and trains women to undertake construction management.
4. Community sanitation designed and developed by Mahila Milan, which creates space for children, women and men for toilets and is managed by women is now the basis for the design of sanitation being developed in Mumbai, Lucknow and other cities.

Uniqueness

The process is sustainable as it involves women centrally from the beginning, creating an agenda for change based on their needs. It moves at a pace they can manage and see solutions, which satisfy them and ensures that they can undertake those solutions on their own. Communities support these processes as women are assisted to negotiate power sharing with men in a manner which is useful for the relationships of men and women and which benefits the family and community. The problem is identified and evaluated to identify its causes and factors restricting in generating the solution. Small pilot projects are undertaken to test the alternatives; resources, which they do not have are supplied by SPARC through development assistance. This demonstrates what is possible, quantifies what women can do, and forms the basis of standardization essential for large-scale solution. Mahila Milan have undertaken these initiatives in house construction, sanitation, etc.

Empowerment of Women

Glenn & Gordon (2000) remark that violence against females between 15 and 44 years old causes more death and disability than cancer, malaria, traffic accidents and even war. And 70 per cent of the world's 1.3 billion poor are female. The survival of the children is related to women's economic power and to their role in society. Improving the status of women could be the most cost effective strategy for addressing most of the challenges we face at the millennium. Sen (1997) mentions that empowerment starts with changes in consciousness and in self-perception. This can be the most explosively created, energy releasing transformation, one from which there is no looking back. Empowerment taps powerful reservoirs of hope and enthusiasm among people used to viewing themselves negatively.

YUVA (Youth for Unity and Voluntary Action)

It is one of the many NGOs working in Mumbai for the rights of the urban poor. It organizes youth and women for social action in housing, health, education and judicial system and offers counselling. YUVA is also active in policy advocacy. It, for instance, provides support to pavement dwellers in Mumbai who are under constant threat of being evicted and of having their makeshift homes bulldozed by the municipal authorities. Often when people are evicted, authorities offer to relocate them to the outskirts of the city, far from their work and from their children's schools. Most of them soon trickle back to their old locations and then the cycle starts again. YUVA educates people about their rights—with respect to housing, employment and schooling for their children. Recognizing that, as elsewhere, most responsibility for household survival falls on women, YUVA also supports such activities as women's savings funds. Similarly, there are other forms of support to the people from initiatives taken by different agencies.

SEWA

The Self Employed Women's Association, (SEWA) was born in 1972 as a trade union of self-employed women. It grew out of the Textile Labour Association (TLA). India's oldest and largest union of textile workers in 1920 by a woman, Anasuya Sarabhai. SEWA's predominant goals are Full Employment and Self Reliance. It believes that poor women's growth, development and employment occurs when they have work, income and food security. It is the members needs and priorities, which shape the needs and priorities of the organization. There are certain questions related to its own operations on which the performance of the organization is monitored. These are related to enhanced employment, income, food and nutrition, health, child care, housing, increase in assets, workers organizational skills, leadership, more self-reliance collectively and individually.

The organization is registered as a trade union under the Indian Trade Unions Act of 1926. The Union is open for membership to self-employed women workers all over India.

SEWA members are workers who have no fixed employee-employer relationship and depend on their own labour for survival. They are poor, illiterate and vulnerable. They barely have any assets or working capital. But they are extremely active economically, contributing very significantly to the economy and society with their labour. In fact 64 per cent of GDP is accounted for by the self-employed of our country. There are three types of self-employed women :

1. Hawkers, vendors and small business women like vegetable, fruit, fish, egg and other vendors of food items, household goods and cloth vendors.
2. Home based workers like weavers, potters, *bidi* and *agarbatti* workers, *papad* rollers, ready made garment workers, women who process agricultural products, and artisans.
3. Manual labourers and service providers like agricultural labourers, construction workers, contract labourers, handcart pullers, head loaders, domestic workers and laundry workers.

The association has 3,18,527 members and membership of Gujarat is 2,05,985.

Membership Pattern

Category of Workers	*No. of Women*	*Prcentage of Total Membership*
Urban	79,008	38.36
Rural	1,26,977	61.64

Membership of Gujarat by Trade

Main categories of Members	*No. of Women*	*Percentage of Total Membership*
Home based workers	72, 156	35.03
Hawkers and vendors	18,759	9.11
Manual labourers and Service providers	1,15,070	55.86

In Gujarat, the SEWA movement comprises of :

1. Co-operatives;
2. Rural Producers (DWCRA) groups;
3. Social Security Organizations; and
4. Savings and Credit Groups.

Federations comprise of arts and crafts, vegetables, savings and credit associations.

At the National Level it has (1) National Centre for Labour, and (2) National Alliance of Street Vendors of India (NASVI)

At the International level, SEWA has presence in South Africa, Yemen and Turkey.

SEWA began organizing workers in the villages in 1979. SEWA believes that the basis of obtaining higher wages is the capacity and power to bargain. However, the workers in these areas had neither the capacity nor the power to bargain because they were weak and vulnerable due to their lack of employment. In a situation where there is an almost unending supply of labour and limited employment, the workers are unable to negotiate for themselves.

Rural Organization has focused on:

1. Increasing employment opportunities for women;

2. Increasing women's bargaining power;
3. Developing women's assets;
4. Capacity building and leadership development of rural women;
5. Providing food and social security;
6. Becoming self-reliant both in economic terms and in terms of running their own economic organizations;
7. Eco regeneration through employment for rural women; and
8. Collaborating with government's rural development programmes.

The role of SEWA can be envisioned by how it worked for the *agarbatti* making workers. There are 20,000 *agarbatti* rollers in Ahmedabad city. About 70 per cent of them are home-based workers and the remainder work in factories. *Agarbatti* workers roll incense sticks for 8 to 10 hours a day so as to make 5,000 sticks. They get their raw materials from contractors or local employers. Women whose husbands or other male family members are laid off from the city textile mills obtain employment in this industry. For the first time ever, thirty *agarbatti* workers held negotiations at SEWA with fifteen of their employers. They demanded that employers contribute 75 per cent of the insurance premium and workers would provide the rest. They also asked for the appropriate work tools, especially a proper worktable to prevent occupational health problems like back strain. The employers have promised to consider these demanas and fixed a follow-up meeting.

Mazdoor Kisan Shakti Sangathan

The case of Mazdoor Kisan Shakti Sangathan (MKSS) illustrates the strength of the collective group. It is more powerful than the legal mandate and can turn around the situation if people collectively stand up for their rights. Aruna Roy quit IAS in 1974 to work with the Social Work Research Centre (SWRC). She set up the Mazdoor Kisan Shakti Sangathan (MKSS), spurring a people's movement for citizen's empowerment and right to information. She was awarded Ramon Magsaysay award for Community Leadership. She was instrumental in launching *Jan Sunvaai*, which is a people's court where people can come together at a place and air their grievances in the presence of officialdom. This gives them an opportunity to know what is being done for them or not being done for them by those who are in power. It is also a forum for public audit to know from the authorities how much was spent. *Jan Sunvai* has given an opportunity to ordinary citizen's to interact with officialdom and people's representatives. The demand for information has a bearing on public ethics, accountability and democracy. Without critical control over these, the poor cannot change the world they live in to bring in a more egalitarian socio-political system.

In the early 1990s the Mazdoor Kisan Shakti Sangathan (MKSS) (Labour Farmers Strength Organization) started working in the Rajsamand district of Rajasthan. The MKSS prepared no project proposals, had no registered society, took no foreign funds and recruited no staff. All they did was walk from village to village asking simple questions – did the people know how much money was coming to their village for development and where is it spent (http://www.barefootcollege.org/html/action.htm). MKSS launched a people's campaign including public hearings of misappropriation and corruption of public funds and a 53-day strike in Jaipur in front of the State Assembly. The strike ended when a gazette of the State Government (a written government order) met the MKSS demands and made them public.

Conclusions

The lessons that can be derived from the analysis of these cases are as elaborated below :

Simple Technologies

Infusion of capital or technology alone is not a solution to the problem of poverty. It is a multifaceted issue and requires interventions at various levels. Merely putting in computers in the villages is not a solution. The reality is that there are large number of people who are illiterate, are unemployed and need to be employed meaningfully so that they could earn their livelihoods. In the cases seen above in various sectors, it has been observed that the technologies which were used were basic, were easy to understand and people have known these for a long time. So the solutions were not imposed from outside but were generated by the communities themselves. The case of the Barefoot College at Tilonia indicates that even the illiterate people can be trained if the technology is easy to understand. Simpler the technology is, higher would be its acceptability by the people.

Documentation of Indigenous Technologies

This points to the understanding of the importance of the indigenous knowledge. Despite technology advances, the problem of drinking water and sanitation is still prevalent. In the olden days, they did not have trained doctors, architects or biologists. They did not depend on theory but applied the wisdom they had acquired over the centuries. Many communities are so fed up with the current state of affairs that they demand that they be left alone to identify and solve their own problems (Roy, 1999). The indigenous knowledge helps to arrive at solutions. Greiner (1998) defines indigenous knowledge (IK) as 'the unique, traditional, local knowledge existing within and developed around the specific conditions of women and men indigenous to a particular geographic area'. Indigenous knowledge can provide insights into the area of food security, health, education, natural ecological initiatives. The advantage of the indigenous knowledge is that it is dynamic in character, as it has the capacity to evolve since people have been using them in the past for surviving through centuries. It needs documentation and stored in systematic manner as it has been passed on from generation to generation. The solution does not lie in inflicting western solutions on communities. Interventions fail to induce people to participate because of the absence of instruments and mechanisms that enable them to use their own knowledge.

The use of traditional knowledge, skills and wisdom promotes active community involvement because people depend more on each other. The use of traditional knowledge demystifies the local technologies that can be the basis for sustainable solutions in the future. The more people understand and try out a technology, the greater the chance of the technology getting accepted. In 1997-98, through the use of centuries old technologies, a total of 12 million litres of rainwater was collected in 100 schools attended by 3000 children at the Barefoot Schools. The cost was a mere US$ 0.10 a litre. The schools have teachers with no qualifications. Over 150 young people from nine States of India have been trained as barefoot solar engineers. They have equipped over 2,000 houses in the Himalayas with solar electricity. The practice could be transferred to other places and situations, but it is essential that several conditions be met (www.unesco.org/most/bpik16.htm). The characteristics of these solutions have to be :

1. The technologies have to be rooted in the ground realities;
2. They have to be sustainable;
3. They have to accommodate huge numbers;
4. They have to trap investments; and
5. Solutions to be Built Around the People Through Their Involvement

People need to be kept in mind for whom the solutions are being generated. Another important aspect that is important is the total involvement of the people at the grassroot levels. No external agent can bring about the change in the community externally. Unless the people stand up for themselves, nothing can change. They would have to be united together to stand up for their rights and there is a need for facilitators rather than consultants. Facilitators should be the people who can identify themselves with issues faced by the people and live with them so as to experience what it means to live a life of deprivation. The solution for any community has to be a sustainable solution that leads to a process, which is self-sustaining. It can only happen if the use of the local know-how is usefully channelised so that it can be sustained in the long run. The case of Tilonia as discussed in the chapter is an example of the same. People need to be sensitized to the change. Acceptance of change is far easier when there is a suitable climate created for the same.

Dissemination of Information to the Poor

One of the main problem of the poor people is the lack of information on what is happening around them. The government starts so many programmes for the poor people but the benefits never reach down to them. The government needs to communicate the information to the poor about the programmes and most important is to monitor the implementation of the same. For example, the much hyped Swarna Jayanti Sahari Yojana and Swarna Jayanti Swarojgar Yojana, which was launched in April 1999, have failed to benefit even 5 per cent of the target population in Orissa. Bamboo crafts persons in Bhubaneshwar which were supposed to be the beneficiaries of the Swarna Jayanti Sahari Yojana was meant to help them rise above the poverty line. For two years, since the scheme began, not a single person received even a rupee as a loan (www.ndtv.com, Marl, 2001), Pro-poor Policies failed because of the lack of reforms. In Khurda district, out of a total of 20,000 families living below the poverty line, only 4,500 were sponsored for loans by the government. The bank sanctioned 1,273 loans out of which only 904 people received the money.

Better Management of Resources

A 200 news reported in the press is an eye opener. The drought affected people in ten districts of Rajasthan protested to highlight that while the Food Corporation of India godowns were bursting with foodgrains, people remained hungry. The FCI had nearly 410 lakh tonnes of food grains in Rajasthan, nearly 139 lakh tonnes in excess costing the government Rs. 4,200,00 lakhs (www.ndtv.com, May 17, 2001). The Food Corporation of India has in recent years been grappling with the problem of how to take care of the huge surplus foodgrain which is fattening ever growing army of rodents or simply rotting in its badly managed warehouses. (*Times of India*, May 11, 2001). So there is a need for a better management of the resources. With little change in the way things are operated, it becomes quite easy. But the things have not changed even on 2005.

Political Will to take Pragmatic Decisions

The income and consumption patterns of the poor are changing. The shift is away from course grains to wheat, paddy and oilseeds although the chemical composition of coarse grains is better than rice and wheat in many cases. Pearl millets have a higher concentration of protein, fat and minerals particularly calcium. Singh (2001) has analyzed the dynamics of the cropping patterns. A drop of 50 per cent in the cultivated area of sorghum, little millet and finger millet has come about in the 1990s. It was in the 1980s that the Public Distribution System (PDS) became a welfare instrument to provide essential items at nearly half the market price. Neither crop loans nor crop insurance are available for these groups. Also there are no subsidies. The promised minimum support price of coarse grains is denied to farmers due to government non-intervention.

The 1999-2000 Union Budget projected a cut in the central deficit of 0.9 per cent of GDP. Achieving this target depended on substantial rise in tax revenue and containing expenditure to only 9 per cent. The interest costs of the debt have increasingly crowded out infrastructure, maintenance and social spending from central and State budgets. Implicit and explicit subsidies at the centre and especially at the State levels are a major factor in the deficit (World Bank Report, 2000). The Ministry of Finance estimated these subsidies at over 14 per cent of GDP in 1994-95. In addition to increasing the deficit, they are distortionary, non-transparent and at best have uncertain equity consequences. At worst, they are anti-equity. Another structural factor is the deficit in the tax system, which declined by over 1.5 per cent of GDP over 1991-98. The tax base, is narrow with 15 million taxpayers.

Transparency

Transparency is required in all operations. A review of literature suggests that poverty has a relationship with political corruption. Oyen (1997) in *Human Development Report* (1997) pp. 95 mentions that poverty often serves the vested interests of the economically powerful, who may depend on the poverty stricken to ensure that their societies run smoothly. A mobile pool of low paid and unorganized workers is useful for doing the "dirty, dangerous and difficult" work that others refuse to do. Corruption in government increases poverty in many ways. Most directly, it diverts resources to the rich people who can afford to pay bribes and away from the poor people who cannot (Transparency International, 1996). Corruption also skews decisions in favour of capital-intensive enterprise and away from labour-intensive activities more likely to benefit the poor. Corruption also weakens the government and lessens their ability to fight poverty. It reduces tax revenues and thus the sources for public services. Most generally, corruption eats away at the fabric of public life – leading to increased lawlessness and undermining social and political stability.

Where corruption is rampant and evidence is in place that development funds have never reached the target segment, it is all the more required. The Barefoot College is evidence where through *Jansunwai*, it shared all the financial details with the people for whom the development work is being done. It must involve participation by the poor if it were to yield meaningful inputs. It is necessary to vitalize the community media with the involvement of people (Sharma, 2000). It is important to bring information to the doorsteps of the poor people.

There is an urgent need to ensure access to modern information technology in rural areas or disadvantaged communities to disseminate simple, practical knowledge, which can save their lives, increase awareness and stimulate development. Properly used, media can help reduce the conflict and strengthen local organization. It can help reduce poverty through providing information on how people of their type somewhere else are handling similar situations. The poor have inadequate access to information, technology, expertise and resources.

Linkages between Different Sectors

Interventions taken in one sector have an impact on the other sectors and there are strong interlinkages among them. Improving the health not only improves well-being but also increases income-earning potential. Education not only improves well-being but also leads to better health outcomes and to higher incomes. Providing protection for poor people not only makes them feel less vulnerable—it also allows them to take advantage of higher risk, higher return opportunities. Increasing poor peoples voice and participation not only addresses their sense of exclusion' it also leads to better targeting of health and education services to meet their needs.

Survival in poverty includes many strategies, which are combined in a process aimed, not just at income in the broadest sense, but also at assurance against the stresses and shocks to which poor people are particularly vulnerable. One such strategy is a mixture of jobs, some temporary, some full time, some self employed, some working for others. Technology could play a vital role in the elimination of illiteracy. For example, the information technology could be utilized to make the education reach to the most distant location. It could create a difference by dissemination of information using various means in rural areas. It could address concerns like the problems faced by craftsmen, farmers or persons, who run a leather tannery, among others. Another important aspect that can be addressed by the IT could be converting entire literature in a language, which the masses understand. But IT alone cannot be a solution. The solution has to be rooted into the reality of the situation.

REFERENCES

Anzorena, E.J. (1994), *Housing the Poor: The Asian Experience*, Philippines:

The Asian Coalition of Housing Rights.

Bhagwati, J. (2001), Growth, Poverty and Reforms, *Economic and Political Weekly*, March 10.

Economist (1999), August 14, "Helping the World's Poorest", pp. 17-20.

Galhardi, Regina, M.A.A. (1995), Biotechnology for Poverty Alleviation in Third World Countries in R. Heeks, P. Bhatt, M. Huq, C. Lewis and A. Shibli (ed.) *Technology and Developing Countries*. London: Frank Cass.

Glenn, J.C. & T.J. Gordon (2000), *State of the Future At The Millennium*, American Council for The United Nations University, Washington.

Gopalakrishnan, R. and A. Sharma (1998), 'Education Guarantee Scheme in Madhya Pradesh', *Economic and Political Weekly*, Vol. 33, No. 39, 26 September.

Grenier, Louise (1998), *Working With Indigenous Knowledge: A Guide for Researchers*, IRDC, 1998

Henderson, H. (1999), *Beyond Globalization: Shaping a Sustaining Global Economy*. Connecticut: Kumarian.

Hindustan Times (2000), 38,000 Primary Teachers' Posts Vacant, May 1.

Human Development Report (1997) *UNDP*, New York: Oxford University Press.

Kondo, E.K. & V. Jauhari (2001), Poverty and Technology – A Missing Link, Working Paper, 2001.

Kothari, R. (1993), *Poverty: Human Consciousness and the Amnesia of Development*, London: Zed Books.

Mazdoor Kisan Shakti Sangathan (MKSS) (www.timesofindia.com, August 7, 2000) Collector to Collectives.

Parayatna Sansthan, Solvta, Dist. Jaipur, Roof Water Collection for the Provision of Safe Drinking Water, (www.unv.org/projects/collect/swrc2.htm). May 10, 2001.

Roy, Bunker (1999), DRA Conference – Capacity and Competence of Communities—Totally Misunderstood, May 23, http://www.mailbase.ac.uk/lists/dra/1999-05/0045.html

Roy, Bunker (2001), *Management of Water Resources at Local Level—The Repair and Maintenance of Hand Pumps*, http://www.unv.org/projects/sl/collect/swrc1.htm

Sambhavna Trust, Bhopal People's Health and Documentation Clinic, www.altindia.net/bhopal/Sambhavna_bhopal.htmwww.altindia.net/bhopal/Sambhavna_bhopal.htm

Sen., Amartya (1999) *Development as Freedom*, New York: Knopf, pp. 87.

Sengupta, N.K. (1992), *Unshackling of Indian Industry*, New Delhi: World Vision.

Sharma, M. (2000), Information Technology for Poverty Reduction, *www.ouhk.edu.hk/cridal/wrapup/discuss/messages/22.html.*

Sharma, Sandipan (2001), Tilonia 'college' wins Children's World Award, *Hindustan Times.com*, April 20.

Singh, Rana Vikram (2001), Grain of Truth, *Down to Earth*, Vol. 9, No. 24, May 15.

SPARC Bombay (2001), Communicating life stories on the Issue of Shelter through the medium of Audio Tapes, www.unv.org/projects/sl/collect/swrc6.htm.

SPARC, Bombay, Communicating life stories on the issue of Shelter through the medium of Audio Tapes, www.unv.org/projects/sl/collect/swrc6.htm

Swaminathan, M. & V. Rawal (1999), *Primary Education for All*, Kirit S. Parikh (ed) India Development Report, 1999-2000, New Delhi: Oxford University Press.

Times of India (2001), "A Shameful Hunger", May 11.

Transparency International (1996), Sharpening the Response against Global Corruption: *Transparency International Global Report,* 1996, Berlin.

United Nations Environment Programme, *Freshwater Pollution*, UNEP/GEMS Environment Library No. 6 (1991).

www.ndtv.com, (2001), Pro-poor Policies let down by the lack of reform, March 1.

3

Sustainable Empowerment of Women

"*Women is the companion of man gifted with equal mental capacities, she has the right to participate in very minutest detail in the activities of man and she has an equal right of freedom and liberty with him. She is entitled to a supreme place in her own sphere of activity as man is in his. This ought to be the natural condition of things and not as a result of learning to read and write*".

—Mahatma Gandhi

GANDHIAN ON WOMEN'S EMPOWERMENT

Indian freedom movement was instrumental in highlighting many hidden facets of Indian society. One of the aspects that immediately come to mind is the status of women, their relationship with the society, their position within their family, various perceptions about their, strengths, importance, work and their contribution to the society. Alternately perceived as a goddess and personification of evil, her position is further complicated because of biased interpretations of religious texts.

During the freedom struggle, men were being sent to prison without trial by the British. Women contributed to the financial upkeep of the family instead of being restricted to the home. The demands of such circumstances forced women to establish small business units at home, so that the familial needs could be met. It caused a major upheaval in their placid homebound lives. And before long women were an enthusiastic participant in a life outside their homes. It is thus not surprising that India has had the unique distinction of having the most progressive constitutions in the world, which guaranteed equal rights for men and women. No Indian women has ever had to struggle for legal right of equality. Since ancient times Indian women enjoyed an equal and special status in the society.

Exceptions were few. Such was her position in society that no religious function could be considered complete without her presence. Her position was further elevated by the likes of Rani Laxmibai, Razia Sultana, Kasturba Gandhi, Sarojini Naidu, etc. Their will power, determination and single-minded belief in their strengths, were instrumental in empowering numerous women who looked at them for guidance and justification.

Gandhiji was a strong believer in the power of women to silently bring about a revolution in the society. He was of the opinion that the inherent strengths of the Indian women if harnessed productively would usher in '*Ram Rajya*'. Women needed education (this aspect is unrelated to literacy) equality and freedom to steer her destiny. According to him, the role of women in the development of the nation was of prime importance. It never ceased to surprise him that in Indian society, a woman could easily transgress various roles with an amazing zeal—a daughter, sister, mother, wife, a queen and a professional and still be committed to a cause. He was a great admirer of ancient Indian women like Sita, Draupadi, etc., because of their moral strength. He felt that these women were not passive and weak; instead, he felt that the true meaning of this resistance was '*Satyagraha*'; the soul force or the truth force, i.e., the power of enlightened non-violence, which was neither passive nor timid.

Gandhiji questioned the blind societal adherence to certain beliefs in Indian society which lowered the status of woman. In his words: "I do not subscribe to the superstition that everything is good because it is ancient..." In his message at the All India Women's Conference (December 23, 1936) he said, "When woman, whom we call *abala* becomes *sabala*, all those who are helpless will become powerful". He further added that such empowerment is to be gained only when women gather strength to stand up on their own or "lift the self, by the self".

Shanties to Bapu more than fifty years since he passed away, Indian women have far greater visibility and voice. They have initiated many movements revolving around issues pertaining to their growth and development. The issue of women empowerment is still at the centre of political and developmental dialogue in India, for there are huge gaps between constitutional guarantees and the ground realities of women's lives. Part of this neglect comes from policy makers' reluctance to deal with issues that are intrinsically linked with the public criteria, and religio-cultural beliefs. Women and girls bear the brunt of the disparities in legal, social and economic rights. Lack of suitable institutional environment, equal opportunities and access and control of the natural resources persistently compound these inequalities.

Background

Empowerment, as expressed by Sen and Batliwala (2000), "Empowerment is the process by which the powerless gain greater control over the circumstances of their lives. It includes both control over resources and over ideology....[includes, in addition to extrinsic control] a growing intrinsic capability—greater self-confidence and an inner transformation of one's consciousness that enables one to overcome external barriers......". The crux is the ability to achieve one's goals. Women's position and degree of empowerment is defined by gender and gender relation in the society. Gender represents not only biological sex of an individual, but also the different roles, rights and obligations that are attached by society to individuals born with male or female sex characteristics. Unequal gender relations imply that men not only have and can exercise greater power than women in almost all spheres of functioning, they also have culturally and often legally sanctioned power over women and have greater control of and access to resources and information. The inequality in gender relations is embodied in various societal institutions, but is reproduced daily in households.

The study on empowerment indicators by Kishore & Gupta (2004) that relative to men, there has been little change in empowerment of women over time. There is greater variation in the level of women's empowerment across different States and regions.

According to Mason & Smith (2003) three key points illustrate the conceptualization and measurement of women's empowerment. The first point is that the power within the household is strongly influenced by social contexts (national and community); the second point is that all aspects of women's empowerment are multi-dimensional, and interrelations among different dimensions depend upon social context; and the third point is that traditional socio-economic proxies are not comprehensive enough to measure the degree of empowerment and, therefore, direct measures should be used. They further add "that studies of women's empowerment need to focus on rights, obligations and resources granted to females *versus* males under different gender systems rather than on characteristics of individual women". In other words, acquiring some standard of financial self-sufficiency and autonomy is a prerequisite to any progression of women at a social, cultural and political level.

According to the estimation of the Regional Labour Commission in 1991 there were about 20 million women in rural areas who were doing home-based work alone and this did not include contract workers. These women were largely excluded from and national statistics because they worked from home, they preferred to cult themselves as housewives during the surveys despite spending 14-15 hours a day earning an income (Sen & Kumar, 2001). Table 3.1 indicates most women are still involved labour intensive activity.

Table 3.1: Sector-wise Distribution of Women Workers

(As per cent of Total Women Workers)

Category	*1981*	*1991*	*2001*
Cultivator	33.1	34.2	32.9
Agricultural labour	46.3	44.9	38.9
Livestock, forestry, etc.	1.8	1.6	N.A.
Mining and quarrying	0.4	0.3	N.A.
Household manufacturing	4.6	3.5	6.5
Other manufacturing	3.6	3.9	N.A.
Construction	0.9	0.7	N.A.
Trade and commerce	2.0	2.3	N.A.
Transport/communication	0.4	0.3	N.A.
Other services	6.9	8.3	21.7
Total	**100.0**	**100.0**	**100.0**

Source: Census of India 1991, p. 55 in Sen & Kumar 2001.

All opinions and debates on the issue of women's empowerment ultimately have to cross the Rubicon of income earning opportunities and their empowerment potential for the women, where the location and work methodology can negotiate an alliance between access to regular income and improvement in women's position. According to Kantor (2002) "moving women out of dependent sub-contracting relations that have fewer sources of work and a great potential for male knowledge of enterprise operations may be a means of improving women's income control. This in turn would also impact on empowerment as measured by involvement in other household decisions as income control and decision making involvement are clearly linked". "The silver lining in the cloudy sky has been the development of several grassroot movements initiated by women themselves, especially from the lower sections of the society" (Ramanathan, 2002).

"Though, the women's movement and a widespread network of non-government organizations which have strong grassroots presence and deep insight into women's concerns have contributed in inspiring initiatives for empowerment of women, there still exists a wide gap between goals enunciated in the constitution legislation, policies, plans programmes, and related mechanisms on one hand, and the situational reality of status of women in India on the other"—[*National Policy for Empowerment of Women* (NPEW), 2001]. In fact, increased use of cheap female labour has become an important strategy for landowners to cut costs in agriculture. At the same time, in the manufacturing sector, women's wages are less than half of men's wages, according to Sen and Kumar (2001) the reason for this segregation of work in this sector is partly because women are engaged in home based work on piece rates and since there is no clear employer-employee relationship in the production chain which involves contractors, wages are usually less. They further state that in several manufacturing industries there is a shift towards, "flexibilization" of the labour force, i.e., employing contract workers instead of full time workers, where women are losing their jobs because of industrial restructuring. The NSSO (1987-88) data in Sen and Kumar (2001), (see Table 3.2) throws up startling results on average wage earning/day by sector of work as far as women's wages are concerned. The Table 3.2 also points at a wide disparity in average wage earnings in almost all the sectors and the pattern is similar in both rural and urban areas.

Table 3.2: Average Wage Earning/Day by Sector of Work

Sector	*Rural*		*Urban*	
	Male	*Female*	*Male*	*Female*
Agricultural	14.58	10.65	27.57	19.05
Mining/Quarrying	38.70	23.81	46.23	31.73
Manufacture (Household)	2.77	9.42	29.83	13.06
Manufacture (Other)	28.69	13.78	43.59	33.20
Electricity	36.16	45.17	50.14	43.93
Construction	31.58	18.27	39.23	27.65
Trade	17.99	18.42	22.55	18.37
Transport	32.08	26.80	41.51	47.20
Services	46.25	38.75	62.10	53.92
All	**30.14**	**21.56**	**41.87**	**34.76**

Source: NSSO 1987-88 in Sen & Kumar, 2001.

Wage disparities are a manifestation of gender inequality that expresses itself in various forms, like the declining female to male ratio, social stereotyping, violence at domestic and societal level, etc. Keeping this aspect of social reality the NPEW (2001) has identified specific objectives, which proposes to address and rectify the prevailing situation.

OBJECTIVES OF NATIONAL POLICY FOR EMPOWERMENT OF WOMEN

1. Creating an environment through positive economic and social policies for full development of women to enable them to realize their full potential;
2. The *de jure* and *de facto* enjoyment of human rights and fundamental freedom by women on equal basis with men in all spheres—political, economic, social, cultural and civil;
3. Equal access to participation and decision making of women in social, political and economic life of the nation;
4. Equal access to women to health care, quality education at all levels, career and vocational guidance, employment, equal remuneration, occupational health and safety, social security and public office, etc.;
5. Strengthening legal systems aimed at eliminating all forms of discrimination against women;
6. Changing societal attitudes and community practices by active participation and involvement of both men and women;
7. Maintaining a gender perspective in the development process;
8. Elimination of discrimination and all forms of violence against women and girl child; and
9. Building and strengthening partnerships with civil society, particularly women's organizations.

Source: National Policy for Women Empowerment (2001).

According to World Bank Policy Research Report (2001), gender equality is the core development issue—a development objective in its own right…. it strengthens countries' abilities to grow, to reduce poverty, and to govern effectively. Promoting gender equality is thus an important part of a development strategy that seeks to enable all people—women and men alike—to escape poverty and improve their standard of living. This report further suggests a three-pronged strategy for promoting gender equality—

1. Reform institutions to establish equal rights and opportunities for women and men;
2. Foster economic development to strengthen incentives for more equal resources and participation; and
3. Take active measures to redress persistent disparities in command over resources and political voice.

It is against this backdrop, that we have analyzed two distinctive case studies. Both of these studies have unique characteristics. The overview about SEWA points at the struggle for equal rights and opportunities for women in a work arena outside their houses. While the study of Lijjat, looks for freedom and emancipation of women inside the four walls of the house. Yet, they converge at a point where their actions begin to impact on their standing inside their households as in case of SEWA and their social standing outside their households as in case of Lijjat.

THE CASE OF SEWA

Self-Employed Women's Association (SEWA) came to its own in 1981, when it broke away from its parent trade union, the Textile Labour Association (TLA). At the time of its formation in 1971, women felt that SEWA should establish itself as a Trade Union. It was an uphill struggle. Finally in 1972, SEWA was registered as a Trade Union. It is an organization of poor, self-employed women workers. Most of them belong to the unorganized zone of the total labour force in India, and form a whopping 94 per cent of the total women labour force employed. SEWA aims at organizing the women who are working as labour or run small businesses, and who have no support for their work or for themselves in times of difficulties. These women do not earn a regular salary or have any welfare benefits as their counterparts in the organized sectors. These women are illiterate, poor and extremely vulnerable. With barely any assets and working capital, they readily fall prey to numerous exploitative forces. Though economically very active, these women are not accounted for in any official labour policy. SEWA organizes these women, namely vegetable/fruit/fish/egg hawkers/ vendors, coolies, casual mill working, women who constantly move between different localities to sell their wares, manual workers and home based workers like weavers, potters, *bidi* and *agarbatti* workers, etc., to accomplish their objective of full employment and self-reliance through the strategy of struggle and development.

The guiding philosophy of SEWA is the Gandhian thought of *Satya* (truth), *Ahimsa* (non-violence), *Sarvadharma* (integrating all faiths, all people), and *Khadi* (propagation of local employment and self-reliance). It functions as a confluence of three movements, namely, the labour movement, the cooperative movement and the women's movement.

"This union is open for membership of self-employed women all over India if they can pay the token fee of Rs. 5 per year. A two-tier level of elected representation governs the union. The members of trade elect their representatives in the ratio of 1 representative per 100 members. These representatives then form the Trade Council (Pratinidhi Mandal). In addition, and parallel to the Trade Councils, are the Trade Committees (Dhanda Samiti) in each trade" (SEWA).

Members of SEWA strongly feel that women's progress, employment and development occur only when they have a regular source of income, a roof over their heads and an assured food security. They have recognized a universal truth that only healthy women can beget and rear healthy children and, therefore, access to women and child health care are a definite prerequisite for development. And so, in order to pursue their twin goals of Self-Reliance and Full Employment, these women developed their own guidelines. This has helped them in evaluation of priorities/needs, monitoring of activities and direction of the organization, so that various activities of their organization remains in conformity with the needs and requirements of its members. These aspects also increase the accountability of members.

Given below are the ten questions that act as evaluatory guidelines for the members, group leaders, executive committee members, and full-time organizers of SEWA.

Another step, which helped these women to further their goal of self-reliance, was the establishment of the SEWA bank in 1974. At the initiative of nearly 4,000 members, this bank was established as a separate bank for the poor self-employed workers. One of the objectives of this institution was to give credit at reasonable rates to the self-employed women. This was done keeping in view the needs of empowerment of these women and to minimize the uncertainty of availability of loans that these women faced while approaching the moneylenders.

Table 3.3 : Ten Self-evaluation Questions of SEWA

1. Have more members obtained more employment?
2. Has their income increased?
3. Have they obtained food and nutrition?
4. Has their health been safeguarded?
5. Have they obtained child-care?
6. Have they obtained or improved their housing?
7. Have their assets increased? (e.g., their own savings, land, house, workspace, tools or work, licences, identity cards, cattle and share in cooperatives; and all in their own name.)
8. Have the workers' organizational strength increased?
9. Has the workers' leadership increased?
10. Have they become self-reliant both collectively and individually?

Source: www.sewa.org/aboutus/goals.htm

The initial capital of the bank was Rs. 60,000. From the very beginning, the bank was professionally run with qualified members hired by the SEWA board. Today with over 60,000 depositors and a working capital of over Rs.100 million, it is a financially viable entity. This bank, besides, strengthening individual workers with its surplus, provides all finance linked supportive services to its members. At the initial stage, the bank used the services of its members to mobilize depositors to get loans from nationalized banks. It later discontinued this scheme, instead it began to advance loans to its depositors, to enable them to own hand-carts, sewing machines, looms, street corner shops, tools of carpentry and blacksmithy. This has encouraged them to upgrade their skills and develop more business

The bank has also made inroads into rural banking and is currently establishing many rural branches. By providing loans to women members for (1) working capital, (2) work tool, and (3) and for housing, this bank has demonstrated that even poor women save and use loans productively. The SEWA bank provides these women an opportunity in financial management and an exposure to innovative money management techniques, so that they can make their money more productive.

They also repay these loans in time. This bank also acts as a facilitator in transferring assets to the women in their names, by insisting that housing loans are in the name of the women and that the house subsequently should be in their names. Often farmers mortgage their agricultural lands in order to tide over financial difficulties. SEWA helps them to recover the land at the same time it also transfers it in the name of the woman of the household as a part of asset building programme. By helping the rural women to build and manage their own water structures in the areas of drought, SEWA has sought to build a relationship of trust and involvement into the lives of the under-privileged women. This has enabled women to shun moneylenders, develop skills of dealing with officials and government organizations. "In the process, their self-confidence has been enhanced. The vicious circle of indebtedness and dependence on middlemen and traders has been broken. This has changed the bargaining position of these women." (www.unesco.org/most/asia1.htm). SEWA also creates visibility for these women with the help of its Video wing (as mentioned earlier in the book), by holding seminars, *yatras* and by lobbying with the government for privileges.

Technological intervention practiced by SEWA is primarily in two areas, namely publicity/awareness and asset building. With the help of the media SEWA has brought to the forefront various difficulties faced by these women as they go about eking their livelihood. It has persuaded numerous organizations to contribute by training these women in health, rural marketing, video film making, and to build and maintain water resources (wells, hand pumps, ponds) with the help of local technology. The fallout of these campaigns have been two fold: (1) external agencies are now convinced of the seriousness of the organization and are thus coming forward to assist it, and (2) the women of SEWA are becoming more aware of their rights and duties towards themselves, their children and the society. Loans on cheaper rates, training and awareness has helped them to create assets for themselves in the form of land, house or employment related tools. It has also emboldened them to send their children to school and to keep their family size restricted and to form cooperatives to explore more business development areas such as export of handicraft and marketing of food and allied products for domestic and foreign market. In other words, according to UNESCO, SEWA has been instrumental in creating assets worth more than Rs. 200 million for more than 60,000 women in India.

The Case of Sri Mahila Griha Udyog (Lijjat Papad)

Seven women, a terrace and an annual ritual of making *papads* for the family.... This was all it took for a 315 crores (including 12 crores exports) industry to take birth.

The year was 1956, the day 15th March, seven middle aged Gujarati house wives had gathered on the terrace of their community building to make *papads* for their families. Yet, it was not one of those routine days, it was a day when they decided to pool their *papad* making skills to augment the income of their families. According to the lone surviving member of the group of seven, Jaswantiben, the *papad* making business of one Laxmidasbhai was running into losses, so they decided to buy him out with a princely sum of Rs. 80. Subsequently, they bought more ingredients and set the process of making *papads* in motion. The business registered itself as a village industry in 1966 and thus Sri Mahila Griha Udyog (SMGU) formally came into its own.

Right from its infancy SMGU, had resolved to be an organization of the women, for the women, and by the women. Since then, only women can become member co-owners; men who wish to join, the organization can only do so in the capacity of an employee. All the members are referred to as 'Bens' or Sisters. SMGU, has demonstrated the concept of trusteeship (by making all its members its co-owners), which was so near to Gandhiji's heart. He felt that the worker should also be the owner of the organization. He/she should have a share in the loss or profit of the organization and it was in their interest to strive for an increasing growth of the product and its market. In other words, he envisaged an organization that grew collectively with equal contribution from the workers. Gandhiji considered trusteeship as a refined form of service wherein all the assets/businesses belong to the society. And in this way the ownership of the business was transferred to everyone engaged in its functioning, that is to say that there were/are no employers or employees.

At SMGU, this idea is emphasized at every meeting/training session/and publication. Not surprisingly, every decision is based on consensus and any member of the organization may veto it. This aspect also empowers the members to take on more responsibilities in order to ensure that their mission is successful and remains profitable.

The organizations gains from this by being able to access dedicated workforce from among its own ranks, which can subsequently be promoted up the hierarchical position. The current

President of SMGU, Ms Jyoti Naik is a case in point. She started her career at the age of ten by rolling out *papad* during the day, and eventually rose to the position of the President.

The organization has also stuck to its philosophy of never accepting any charity or donation. The fallout of this can be seen in the immense confidence with which these women retain and guard the independence of their organization from interference from outside agencies to influence its policies. This also meant that they had to do away with the notions of pity, sympathy and charity, even though each of its members comes from the poor sections of the society. Independence of the organization has inculcated in them a sense of pride and capability to earn their livelihood with their own efforts independently. Increasing confidence in their own abilities have encouraged these women to give back to the community a small share of their earnings either in cash or kind. The organization facilitates easy access to health care, scholarships, and house constructions for its members who often utilize its services to promote the welfare of their families. It has also set up numerous educational hobby centres to help rural women acquire new skills. It also conducts courses in first aid, hygiene, and childcare. The members of the organization have contributed in various ways to many cleanliness drives, rural reconstruction, water and fodder distribution and tree plantation efforts in and around their respective branches and villages.

Work and regular income has also increased the importance of literacy among the women. They are no longer reluctant to send their children, especially the girls, to school and for higher education. This aspect is further facilitated by the provision for scholarship and monetary help, which is provided to them by the organization. Another aspect of education is the introduction of computer education for the daughters and sons of the members. Besides educating their children, these women themselves have become enthusiastic in becoming literate and, therefore, many branches run literacy classes for half an hour every morning when these women come to hand over their day's production of *papads*. Their motto being that women cannot be totally independent, unless she is literate. Their teachers were most often educated daughters of the other members.

How does Lijjat go about its Business?

The details of the operation are explained in the Figure 3.1. Every woman who is willing to work in any capacity, irrespective of caste or religion, can become a member owner of the organization. All they have to do is to sign a pledge of devotion to the basic principles of the organization, which is a 'commitment to earn *legitimate* honest income through work on cooperative basis' (Ramanathan, 2000). Girls can join the organization at the age of 18 and can leave the organization whenever they wish. On its own, the organization does not ask any one to leave, unless she violates the tenets of the organization. Currently the organization has over 40,000 women employed as co-owners in 40 divisions across India (*Times News Network*, 2003).

The Central Office of the organization in Mumbai dispatches all the raw material to its branches (62 in all, spread over from Jammu and Kashmir to Goa and beyond). It also conducts business with the exporters and suppliers. Besides, it audits all the accounts sent to it by all the branches, which have to maintain them on a daily basis.

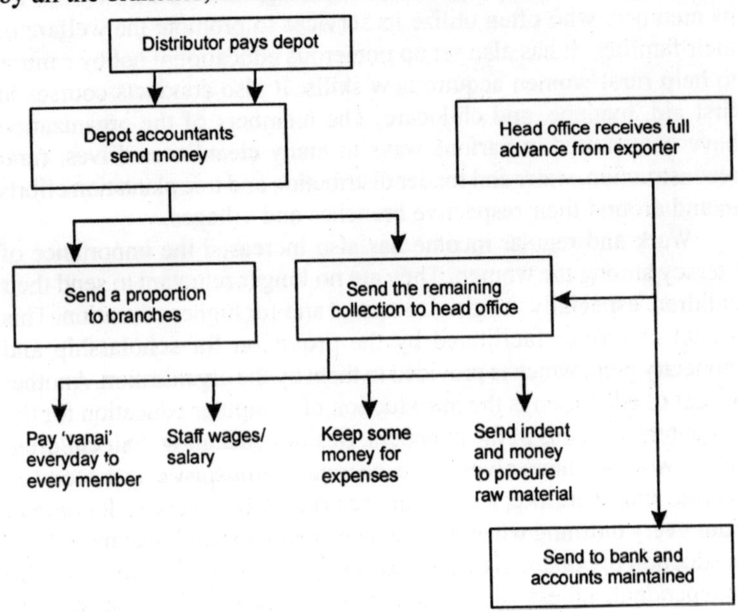

Figure 3.1 : Collection Flow Chart

Source: *The Smart Manager,* April-June 2003, Vol. 2, issue 2.

A managing committee comprising of 22 members, including the president, vice-president, two secretaries and two treasurers have been entrusted with the responsibility of running the organization. Every three years, the members of the organization choose the office bearers with consensus by carefully scrutinizing their contribution and leadership qualities. In order to be considered for the office, a member has to express her wish in writing. A Sanchalika selected by members every three years heads each branch. They can be transferred to any branch, which has a committee of eleven members, again chosen by consensus. "Each branch acts as a self-administering unit, sharing its profits and losses among its members" (Ramanathan, 2000). Whenever a new branch opens, members from the nearby area train the new members who join the new branch. And no sooner are the new members confident enough to handle the affairs of the branch, they are entrusted with the administration of the branch. The central office has the overall responsibility of these branches.

"Each morning a group goes to the branch to knead dough; this is then collected by another set of women who take it home to roll it into *papads*. When these women come in to collect the dough, they also give in the previous day's production, which is tested for quality. Yet, another team packs the tested papads. Every member gets her share of '*vanai*' (rolling charge) every day for the work she does (Naik, 2003).

SMGU has also diversified into making other products besides *papad*. In order to maintain quality and consistency of the raw materials, the Khakra division was set up in 1974; this was followed by Flourmills in 1975, and the *Masala* division in 1976. SMGU also has its own printing division set up in 1977 and a Polypropylene Packing Division for packing the finished products. Lijjat diversified into making Detergent Powder in 1988 to make synthetic detergents under the brand name SASA, which are cheaper and efficient than ordinary soaps. This industry has shown a consistent 15 per cent growth per annum since 1983 (Ramanathan, 2000).

Quality has always been the USP of Lijjat. No sooner does a member joins the organization, she is constantly reminded to make quality her maxim. She is trained to make a perfect *papad*. So distinctive and complete is the training/indoctrination in quality that, every consumer, wherever he/she is gets the same consistent quality of *papad*. According to Jyoti Naik, President of Lijjat, "Our quality does not differ whether it is for exports or for local market. There is just one quality. And that's good quality. Again and again and again".

Conclusion

Women empowerment is very critical for moving the society to a stage of development. Women with their unique skill sets can do multiple tasks with lot of dexterity. A literate woman can be a great support to her family and contribute immensely to economic growth. There is a change of mindset specially in bigger cities but in rural areas interventions such as Lijjat and SEWA are still scarce. A large number of villages still have majority of women who have to struggle to get water, do household chores and work in fields to make both ends meet. There are disparities in income, opportunity, education and treatment meted out to them in social conditions. The law in principle ensures equality but the societal culture has a huge bearing on how she is actually treated at the workplace. The other side is that how many women really take their work seriously and how much support is extended to her to work equally as a man. All these issues need to be addressed in order to create workable options for women. There are numerous instances that have been witnessed and feminist movements that have gained ground. The term "liberalised woman" has been misconstrued. Liberalisation is not about working late hours, smoking, drinking and being loose in your relationships. Liberation is a quality of mind. The sense of self-esteem, self-respect, power to take independent and considered decisions rather than proving a point, ability to develop ones skills and generating opportunity to find avenues for expression of talents is true liberation of mind.

Meeting or attempting to rise to standards of success as laid down by a society is itself being slave to a particular mindset. Always struggling to meet somebody's expectation is a false claim to liberalization. The mind needs to be pure, at peace and truly unchained and in a state of bliss – away from wanting and accumulating more, generating a sense of false pride.

The above discussion is at a philosophical level. To actually relate it with reality, there is a role which different stakeholders have to play. The government, society, women, NGOs international organizations have a role to play in bringing about a change. To bring in a change, some of the following initiatives can be undertaken:

Information Dissemination

When someone has lived in deprivation and with the feeling of an inferior status for too long, then it is essential that there is a need to work on that mindset. There is a need to infuse confidence and an awareness of abilities of an individual. Here extensive government campaigns, media, drama, NGOs, can play a major role. Media can play an immense role. The stories, art, theatre and movies reflect a culture at a particular point of time. It is important that positive role be played and influence be cast. Women's role can be portrayed in a positive manner.

This realization that a woman is powerful and has a legal support can change the situation. There may be an eventuality that it could be misused as has been the case such as dowry where non-bailable warrants can be used to check the growing dowry cases. However, care needs to be taken to avoid such an abuse of power.

Single Point of Contact

For a woman to make an economic contribution, it is essential that support mechanisms must be in place. When a woman with a family, goes to work, the children need to be taken care of, the food needs to be prepared, and other household chores need to be addressed. Such issues are very real. In a situation where men and women roles are so clearly defined despite literacy, it becomes doubly hard for women to cope with work pressures. The endeavours such as those by SEWA address such issues. The community provides day care centres, health care and education as well for children. Empowerment of women will not have any meaning if the issues are not addressed in a holistic manner.

Entrepreneurial endeavours such as Lijjat and SEWA are required. In rural areas, if women come forward and start vocational centres or education with others volunteering to manage other things such as community food, day care, washing, it could lead to prosperity on the whole and make life more liveable. The Barefoot initiative as mentioned in chapter two discusses the education intervention as initiated by Children Parliament as immensely successful.

There is a need for single point of contact for women. For example someone wants to set up an entrepreneurship venture and wants seed capital, then information needs to reach the women with minimum hassle. These information centres need to be established at the village level. The bank loans, child support, food and other related aspects all need to be tied up to make such an entrepreneurship possible. Here then the onus of such an intervention lies with the community as well. Interest groups must connect with each other to make such ventures successful. Today when someone needs to start up a venture, there are numerous agencies and points where one needs to go to make a beginning. There are bureaucratic hurdles, lack of awareness and many barriers in between the entrepreneurial spirit and venture capital availability. Each village or area has certain strengths which need to be managed professionally.

Each area or unit has to become a centre of excellence, equipped with resources to make it liveable. That would ensure minimal migration and incentive for people to contribute towards its growth. Very often we talk about comparisons with East Asian countries' prosperity. Nation states with size of an Indian metro have been generating more exports than the entire country such as India. India needs to take some lessons from here. We need to look at each village as a centre of excellence and self-sufficient if the entire nation has to progress.

Role Played by the Government

The government has an immense role to play. It needs to allocate funds to let the campaigns gain ground at the grassroot levels for self-reliance. The venture capital needs to be available at the village level. Also, it can identify areas wherein women entrepreneurs can contribute. The mapping of geographical strengths and nature of entrepreneurship that can be sustained needs to be generated. If such projects are in place, then they can be shared, and it will also get the venture capitalists at the grassroot levels.

Kurmanoth (2004) has documented the initiative of HUDA in enhancing women empowerment.

> "About 5,170 women from 595 women groups have earned Rs. 1.25 crore raising to 1.59 crore saplings for HUDA's beautification projects covering 18,688 hectares in last eight years. The women have also diversified their activities. They are now into vermi composting vegetable cultivation, candle making and floriculture. The poorest of poor women are selected for

homestead nurseries. HUDA has also decided on a plan to support prospective entrepreneurs. In the first phase, they will disburse Rs. 6-7 lakhs for about 10 groups after going through project reports."

The International Organisations can also pitch in and adopt new approaches. A case in point is the approach adopted by Asian Development Bank in Bangladesh (World Sources Online, 2004):

> Anticipating that women in Bangladesh would face new difficulties including unemployment after expiry of Multi Fibre Arrangement (MFA) at the end of 2004 as the majority of textile workers are females, the bank is providing support to facilitate women's participation in various areas including grassroots level organisations, self-employment, rural for them economy and local governance. Women's participation in urban governance has been supported by establishing mechanism to take part in municipal level decision making process, citizens committees, infrastructure management and capacity building of the women ward commissioners and other activities under Urban Governance and Infrastructure Improvement Project.

Society

Women empowerment is an issue across various societies in various countries, not only in an organisational context. Adler (1993) states that "The single most incontrovertible statement to make about women in management is that there are very few of them". Opportunity 2000 was launched in October 1991 in UK with an aim of increasing opportunities for women to enter management position (Morley, 1993). The biggest challenge is to nurture confidence in self and also initially to create a climate of nurturance. Once the confidence in self develops, the individual will get affected to a lesser extent by the interpersonal stimuli.

India is one of the world's largest producer of fruits and vegetables. India is also the largest producer of milk in the world. With these as the resources, why do we not have international brands in food processing? This is the reality at the grassroot level. Why so much of raw material gets spoilt in warehouses and transportation.

Why can't we sell the packaged food? Indian food is sought after internationally. Why can't our food flavours be packaged and why can't they be sourced at the village level if suitable technology is instituted. The culinary tradition is deep rooted with Indian women. If one travels through interiors of U.P., Bihar, Rajasthan, Bengal, there are so many rich Indian dishes which could be sourced from these areas and women could be the key drivers for the businesses related with these.

The mindset will also have to be changed from being job seekers to job givers. The entrepreneurial spirit has to be awakened. Also the feeling of being very satisfied with short term gains have to be got over. When right things are done?

Literacy

Here there is a scope for government aid, NGOs, international organizations, volunteers who may pitch in to make the drive for literacy a success. Women need to create self help groups to take this drive forward. The theory of more hands to have a better livelihood will not work in the long run. Literacy has a relationship with prosperity, health and openness of mind. However, reaching in the interiors of the country is immensely challenging.

Micro Credit Availability

Creation of entrepreneurship requires that financial support be extended to the people at the grassroot level. The schemes must be implemented with minimal effort. Deadlines for clearance and support must be extended in conducting feasibility studies at the grassroot level.

REFERENCES

Adler, N., "An International Perspective on the barriers to the advancement of Women Managers", *Applied Psychology: An International Reviw, Sp. Issue, Women in Management*, Vol. 42, No. 4, 1993, pp. 289-300.

Government of India (2001), *National Policy for Empowerment of Women* (2001) cited on 29th November 2003, available from URLhttp://www.wcd.nic.in/empwomen.htm.

Kantor, Paula (2002), *Women's Empowerment Through Home Based Work: Evidence from India*, 66th Annual Meeting Midwest Economics Association, Chicago, Illinois March 14-16, 2002, Cited on 29th November, 2003 available from URL<http://www.uwlax.edu/ba/eco/MIAFFE/mea_2002.htm>.

Kishore, S. & Gupta, K. (2004), Women's Empowerment in India and its States, *Economic and Political Weekly*, Feb. 14, pp. 694-712.

Kurmanath, K.V. (2004), HUDA's Green Drive Shelters Poor Women, *Business Line*, February 18, pp. 1.

Morley, Louise (1993), Empowering Women Managers in the Public Sector, *Women in Management Review*, Vol. 8, Issue. 7, pp. 26.

Mason, Karen Oppenheim & Smith, Herbert L. (2003), *Women's Empowerment and Social Context: Results from Five Asian Countries*, cited on 29th November 2003 Available form URLhttp://poverty.worldbank.org/files/13323_womens_empowerment.pdf.

Naik, Jyoti (2003), *Lijjat*: *The Smart Manager,* Vol. 2, issue 2, April-June, pp. 51-57.

Ramanathan, Malathi (2002), *Grassroots Development in Women's Empowerment in India*: *Case Study of Shri Mahila Griha Udyog Lijjat Papad (1959-2000)*, cited on 29th November, 2003 available from URL<http://www.pcr.uu.se/conferenses/myrdal/pdf/MalathiRamanathan.pdf>.

Sen, Gita & Battiliwala, Srilatha (1997), *Empowering Women for Reproduction Rights'* in Harriet and Presser, Gita, Sen (eds.) Women's Empowerment *and Demographic Processes*, New York, Oxford University Press.

Sen, Kalyani, Menon & Kumar, Siva, A.K (2001), *Women in India*: *How Free? How Equal?* New Delhi, UNDAF.

SEWA(online) (cited 29th November 2003) Available from URLhttp://www.sewa.org/aboutus/goals.htm.

The World Bank (2001) Engendering Development: Through Gender Equality in Rights, Resources, and Voice (summary), Washington D.C.

Times News Network (2002) ' For secret seven, its Question of Lijjat' cited on 21th August, 2002 available from URL <http://www.economictimes.indiatimes.com/cms.dll/html>

World Sources Online (2004), ADB adopts new approach for women empowerment in Bangladesh, Xinhua, World Source Online, London, Nov. 23.

4

Towards Sustainable Livelihood

"*Every agricultural country requires a supplementary industry to enable the peasants to utilize spare hours*".

—Mahatma Gandhi

INTRODUCTION

Gandhian philosophy encompassed the ancient wisdom of India, wherein being poor meant being deprived of material possessions, education and health care. Gandhi sought an in-depth compassionate holistic approach to poverty alleviation and equitable distribution of wealth. He observed that small holding farmers are caught in a situation of low return per unit of productive assets, inaccessibility to resources and market, ignorance and other economic and social compulsions (Belavadi & Niyogi, 1999). He believed that the collective strength of these farmers could ensure bright prospects for the future development of village. If systems could be developed to provide assured market at remunerative prices for producers at the village level, which could also act as a channel to market the production at the same time and not disturb the agro-based economic structure of the rural folk, it could most certainly lead to a developed society characterized by equitable and sustainable livelihood.

It is in this context that the dairy industry occupies a special position among other sectors of agriculture. Milk produced everyday gives a regular income to the numerous small producers. Being highly labour-intensive, it provides a lot of employment.

Milk—also known as white gold—is very versatile and can be used to make an enormous variety of high quality products. The high cost of milk as a raw material has necessitated a high-tech processing industry. Across the world, the whirlwind of changes are also exercising a growing influence on the dairy industry. The number of milk producers is falling rapidly. The dairy processing industry is becoming more and more highly concentrated, the international dairy markets are increasingly liberalized and are giving greater opportunities to low cost producers, including many from developing countries (Schelhaas, 1999). The factor, which distinguishes the dairy industry within agriculture as a whole, is the socio-economic position of dairy farmers. The vast majority of them are small-scale producers, with a weak and vulnerable position in the market; the nature of the business (involving a high percentage of fixed costs) means that they are only able to adjust to market changes in a limited and gradual way. For that reason, in many countries the dairy industry is considered highly important for the livability of rural areas.

Intensive research into the viability of the dairy industry has compelled countries into conceding that any social interaction must be characterized by the following aspects to make it sustainable:

1. *Financial Viability*: It must have the capacity to go on its own. The project must have the capacity to be financially viable and generate enough funds to carry it through. Any effort, which does not generate funds at some point of time, would have to be dependent on external funding which leaves its future in doubt. The Gandhian philosophy also emphasizes on the importance of self-reliance. Which in turn leads to self-respect, ability to take decisions and prioritize things as per requirements. In the absence of same, there would always be compromises, and
2. *Ability to replicate*: Any experiment would have a lasting impact if it has the capacity for being replicated. The Kaira District Co-operative Milk Producers Union is capable of being replicated. For instance when National Dairy Development Board of India was attempting to promote co-operatives in developing India's dairying, it saw an opportunity that could be used for extending dairying on Anand pattern in India. In 1969, the NDDB formulated a programme called 'Operation Flood' which proposed the utilization of the 'food aid' available in the form of dairy commodities from donor agencies such as World Food

Programme to generate funds for replication of Anand pattern dairying in the country.

Operation Flood

The Operation Flood Programme initiatives included the following :

1. Promoting and assisting milk producers in establishment of village dairy co-operatives and co-operative unions.
2. Creation of additional milk chilling, processing, distributing and marketing facilities to be owned and operated by the Dairy Co-operative Unions.
3. Creation of sufficient conversion capacities with the co-operative dairies for balancing the lean and flush supplies.
4. Establishment of a network of storage and long distance milk transport facilities to enable operation of a national milk grid.
5. Equipping the dairy co-operatives with facilities for providing a set of milk production enhancement inputs to farmer members to improve productivity of their dairy animals.
6. Development and provisions of support services such as information systems, training and manpower services for the co-operatives.
7. Establishment and management of centralized support services such as animal disease diagnostic centres, vaccine and biological production, and delivery system to enhance the productivity of dairy animals.
8. Promotion of indigenous development of dairy processing and conservation methods and establishment of dairy machinery and equipment manufacturing facilities in the country in order to cater to the needs of a growing dairy industry.
9. Establishing facilities and undertaking research and development activities.

Operation Flood has been one of the largest dairy development programmes in the world. It has also been responsible for setting up extensive and most modern facilities in milk storage and transport because the special nature of milk (perishable and bulky) leads to the necessity of strict and comprehensive quality regulation and to high transport costs. The initiative also linked India's strengths, which could be channelized to turn the situation to India's advantage.

Under Operation Flood Programme, the country witnessed a phenomenal increase in milk production from around 23MT/annum in 1970s to around 69 MT/annum in 1996-97. India today is the lowest cost producer of per litre of milk in the world, at 27 cents compared to the US (63 cents) and Japan ($ 2.8). India with 134 million cows and 125 million buffaloes has the largest population of cattle in the world. The total cattle population as on October 2000 stood at 313 million (www.aavinmilk.com/dairyprofile.html). Across India, there are around 70,000 village dairy co-operatives. These co-operative societies are federated into 170 district milk producers unions, which in turn have 22 State dairy co-operatives under its supervision. Milk production gives employment to more than 72 million dairy farmers. They have made India one of the leading producers of milk followed by USA. According to FAO estimates, while world production declined by 2 per cent in the last three years, Indian production has increased by 4 per cent.

Table 4.1: Production of Milk in India

Year	*Production in MT*
1988-89	48.4
1989-90	51.4
1990-91	53.7
1991-92	56.3
1992-93	58.6
1993-94	61.2
1994-95	63.5
1995-96	65.0
1996-97	68.5
1997-98	70.8
1998-99	74.4
1999-2000	78.1
2000-01	81.0

Source: www.aavinmilk.com/dairyprofile.html.

Milk production in India accounts for more than 13 per cent of the total world output and 57 per cent of total Asia's production. The top five milk-producing nations in the world are India, USA, Russia, Germany and France.

Table 4.2: World's Major Milk Producers

Country	*1997-98*	*1998-99 (Approx)*
India	71	74.5
USA	71	71.0
Russia	34	33.0
Germany	27	27.0
France	14	24.0
Pakistan	21	22.0
Brazil	21	27.0
UK	14	14.0
Ukraine	15	14.0
Poland	12	12.0
New Zealand	11	12.0
Netherlands	11	11.0
Italy	10	10.0
Australia	9	10.0

Source: www.aavinmilk.com/dairyprofile.html.

Launched in 1970, Operation Flood has modernized India's dairy sector and has created a strong network for procurement, processing and distribution of milk by the co-operative sector. Per capita availability of milk has increased from 132 gm/day in 1950 to 220 gm/day in 1998. The main objective of the operation was to organize dairy co-operatives in the milkshed areas of the village and link them to the metro cities, which are the main market for milk. The efforts undertaken by NDDB have not only led to enhanced production, improvements in the methods of processing and development of strong marketing networks, but also led to the emergence of dairy as an important source of employment and income generation in the rural areas. It has also resulted in improvement in yields, longer gestation periods, shorter calving intervals, etc., through the use of modern breeding technology/techniques. Establishment of milk collection centres has enhanced life of raw milk and enabled minimization of wastage due to spoilage of milk. The fallout of this accomplishment is the ever growing dependence of milk producers on the dairy processing industry, resulting in a strengthened position of the co-operatives in milk marketing and in the processing industry. Many nations have replicated this model of white revolution, solely on the basis of its success in India.

This chapter discusses some of the case studies, which provide evidence towards sustainable livelihoods.

Kaira District Cooperative Milk Producers Union (AMUL)

Long before cooperatives came into picture, dairy farming was a profitable business for contractors, middlemen and traders of milk. In the year 1946, farmers of Kaira District of Gujarat, in order to augment their income sold their entire milk production to middlemen and contractors for a paltry sum. They in turn sold the collected milk to Polsons a private dairy firm engaged in supplying milk to Bombay and its surrounding areas. The customers were satisfied with their milk supply. The contractors, middlemen and Polsons were satisfied with the profit they made.

The only person who did not benefit from supply chain was the poor farmer. Dairy farming provided a supplementary employment and an additional source of income to these small and marginal farmers and, hence, they invested large amounts of their scarce resources in animal feed and fodder, all the while putting in intensive hard work in the care of the milch cattle. Though they felt exploited and cheated, they felt helpless in the face of the existing situation. Until one day, they decide to approach Sardar Vallabhbhai Patel, for advice. Known for his strong views on self-reliance and community spirit, he advised them to "throw out Polson and his contractors". He went on to suggest that they control their resources and with it their destiny by forming cooperatives and marketing their own produce.

Following his advice they stopped selling milk to Polsons and formed their own dairy unit in 1946 at Anand village. This unit was called Kaira District Cooperative Milk Producers' Union (AMUL). The farmers were guided in this endeavour by Mr. Morarji Desai (he was deputed by Sardar Patel to oversee the formation of milk cooperative).

This Union was formed with "a clear goal to ensure that each member received the highest possible share of the consumer rupee". The future course of action of this newly formed Union, was thus based on Gandhiji's philosophy of "production by masses, and not mass production. It was a humble beginning. Only a handful of producers supplied milk to the cooperative. The per day procurement was a meagre 250 litres. Gradually as the advantages of the cooperative percolated to other producers, the number of farmers supplying milk to the union increased to 400 and with it increased the procurement of milk to 5,000 litres per day. (www.Indiadairy.com/ind_profile_factsheet.html).

As is with every success story, soon the fledgling milk cooperative faced its first hurdle. The government of Bombay refused to buy the milk collected by the cooperative. The cooperatives retaliated by suspending the entire supply from Kaira district to Bombay for 15 days. This resulted in the government conceding to the farmers' demands to supply milk through the cooperatives. No sooner did the farmers cross this hurdle, another cropped up. This time it had to do with excess supply of milk during the winter months.

Bombay milk scheme to which the cooperative supplied milk could not absorb the increased winter milk production. The farmers, therefore, had to sell their surplus milk to the middlemen and contractors. This was a setback to the cooperatives capabilities. A proposal to set up a butter and milk powder plant was mooted by the chairman Sri Tribhuvandas Patel. Set up with the financial assistance of rupees 5 million from UNICEF, Government of New Zealand and technical assistance of FAO, this plant became fully operational in 1955. The plant was expanded, to accommodate the rapid growth of cooperative movement, to manufacture sweetened condensed milk in 1958. It was in 1960 that for the first time in the world buffalo milk was used to produce nearly 2,500 tonnes of baby food (based on a formula developed by Central Food Technological Research Institute, CFTRI, Mysore) and 600 tonnes of cheese in India, when in 1960 the new wing of the factory was initialized at Anand.

Kaira Milk Cooperative then commissioned a project donated by OXFAM, to manufacture balanced cattle feed. It also set up additional milk drying capacity in order to meet the defence needs of the country in 1963. This was followed by a new dairy capable of producing 40 tonnes of milk and 20 tonnes of butter daily. In order to consider the growing demands of the consumers for value added products the cooperative set up the Mogar Complex to produce high protein weaning food, chocolate and malted food. (http://www.indiaprofile.com/businessandindustry/whiterevolution.htm)

The Structure of the Cooperative Union

The structure of the Union is indicated in Figure 4.1. Besides the efforts and hard work that went into making the cooperative successful, an aspect which can never be overlooked, and which has since been replicated innumerable times is the structure of the Anand Pattern of dairy development. At the grassroot level are the primary milk producers, with each of them holding a minimum of Rs. 10 worth of the cooperative share. This holding makes the farmer eligible to become the member of the cooperative. These members then elect a managing committee, which then goes on to elect the chairman. Incidentally, all these posts are honorary. These people then employ people on salaries to collect, and test milk. They are also trained in artificial insemination and general administration.

The milk producers are paid in cash in the morning for the milk delivered previous evening and in the evening for the milk delivered in the morning. Each village dairy cooperative is linked to the district milk union, which usually owns and operates milk processing plants and other facilities needed to support the milk producers to boost their output. The profit paradigm is very simple. These cooperatives earn profits for their sale of milk to the Union, a part of these earnings are distributed to each member every year as a form of bonus. The remainder is spent on various socio-economic and social activities of the village. A Board comprising chairpersons of various village societies manages this Union. The Managing Director then selects professionals and supporting staff for the efficient operations of the plant and the Union. In 1974, an umbrella body "the Gujarat Cooperative Milk Marketing Federation Limited" was set up at Anand to oversee the marketing of milk and milk products.

At the village and at the Union level, the cooperatives have created numerous job opportunities. The unemployed youths and the landless farmers of the district are now gainfully employed in the district. This has put an end to the migration of people to the cities. The Union also trains and educates member producers, management committee members and dairy cooperative staff of the village. Operation Flood, in collaboration with NGOs has established more than 6,000 women's dairy co-operative societies, which have empowered women to meet collectively and explicitly make their own decisions. Leadership talents have emerged and self-confidence has grown. It has helped them to be economically self-reliant.

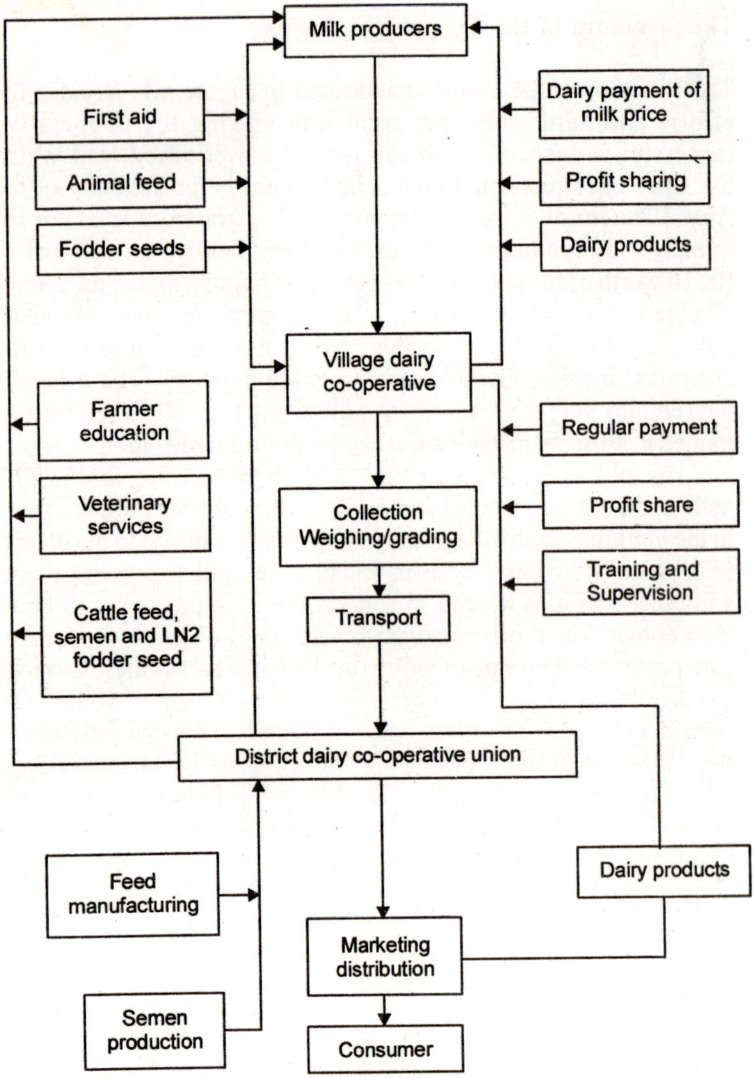

Figure 4.1: Structure of the Union

Source: Belavadi and Niyogi, 1999.

Conclusion

Many great changes, which are affecting the world as a whole, are likewise leaving their mark on the dairy industry. The forces generated by these changes, such as the fast pace of technological progress, economic liberalization, privatization, scale enlargement, internationalization and globalization, are also exercising a growing influence on the dairy industry (Schelhaas, 1999). It is against this backdrop that the Amul story impresses because of its response to the exploitation of the farmer, producer and customer sensitization, market growth and high technological use in dairy products handling. It was perhaps also the first experiment of its kind, which responded to the financial needs of the farmers. It became a commercial success because it had the farmer as its primary stakeholder. It helped millions of farmers to sustain their livelihood by providing alternative gainful employment and necessary technological upgradation. In other words, Kaira was at the vanguard of a grassroot movement, which created a climate for long-term growth in procurement, and equitable distribution of wealth through consumer processing and marketing facilities. It has improved the quality of the farmers' lives with technical inputs like animal insemination, better fodder varieties, emergency veterinary health services, etc. It has brought into its fold landless, small and/or marginal farmers who contribute their milk to the cooperative and thus avail regular cash flow with the help of unexploitative marketing. What had once been a traditional cottage activity, has now transformed into a major economic activity for them. The milk produced and sold by the farmers brought Rs. 320 million a day, or more than Rs. 10 trillion a year (www.1upinfo.com/country-guide-study/india/india141.html).

This, in turn, has influenced the rural financial and economic transactions by changing the consumption pattern of goods and services. The farmer gradually has come to understand that they, and not fate controls their destiny. Income from dairy helps these farmers sustain themselves during adverse conditions such as drought. Socially, these cooperatives have inculcated a spirit of democracy in the members to approach their problems and crises in a consensual manner. The voluntary and willing initiatives of the grassroot members of the cooperatives in managing their own enterprise in a democratic manner has encouraged the development of numerous micro-infrastructural facilities in the rural areas. They are more aware of their rights and duties. Regular interaction of members belonging to different cross-section of the society at the collection centre has promoted a sense of social harmony and brotherhood. Exposure to cutting edge technologies in milk management has had an educative effect on the members.

The once two-village initiative has now transformed itself into a federated body wielding a considerable bargaining power. Modern Management Techniques, professionalism and absolute transparency combined with value added quality products and superior services have made it extremely efficient and competent to offer intense competition to various MNCs. At present, the Union handles approximately 736,000 litres of milk per day. It has approximately 5,553,000 members belonging to 970 village dairy cooperatives (Belavadi & Niyogi, 1999).

Its facilities are capable of handling about 1.5 million litres of liquid milk per day. Amul has demonstrated that illiterate farmers can manage a world class organization, employ highly qualified professionals, utilize modern technology, and exploit market forces to its advantage, and all this without unsettling the time-honoured agro-based system of the villages.

The case of development of Amul as a national brand is an outstanding example of how the local strengths can be harnessed and positioned. There are lot of local strengths that can be channelised and create prosperity in the communities that participate in such a process. There are no short cuts in such a process. Communities have to work hard and take a sense of ownership to change their fortunes. No government plan can ever be successful unless there is a drive by the communities themselves to do well.

Another lesson that can be learnt from such an intervention is that there has been a visionary behind such an endeavour. It has been a foresight of one individual who channelised the energies of thousands of people in one direction. He unified them irrespective of age and caste. However, even bigger than that has been the process of setting systems in place. Many a time, with the leader having gone, the system collapses as systems are not institutionalized. The beauty of Amul intervention is that the systems were set in place. The processes were institutionalized and the legacy continues. The brand Amul is so well entrenched in Indian market that it gives even the multinationals such as Nestle and Britannia a run for their money. The advertisements have also left indelible impressions on the mindset of individuals. The linking of the character associated with Amul with contemporary developments in the market makes it so vibrant. Amul is a wonderful amalgamation of community based intervention with a mix of indigenous strength and technology to harness the local strength.

The intervention also reinforces that entrepreneurial ideas that bind communities can be immensely successful provided community members are stakeholders. When there is a realization among people that initiatives undertaken will help to transform their lives, they get committed to them. It brings out the best in them.

Sustainable livelihood may not be a possibility of individual endeavours. It may more be a reality if there is a possibility of entire community coming together. In such a scenario, different roles may be envisaged for different members. The meaning of sustainable livelihood is not about individual livelihood alone but utilisation of resources with minimal damage to environment which is sustainable in the long run, financially viable and does not collapse after the initiator of the movement has left. The literature does document a debate on the concept. Carney (2002) and Toner (2003) has talked about the underpinning ethos which could be split into normative and operational principles.

Normative Principles

1. People centred: Responding to people rather than resources or services.
2. Participatory and responsive: Processes to listen and respond to poor.
3. Sustainable in all four dimensions: Economic, institutional, environmental, social.
4. Empowering: Increased voice and well-being for the poor.

Operational Principles

1. Conducted in partnership: Formed with poor people as well as public and private sector.
2. Multi-level and holistic: Primarily the micro informs the macro level.
3. Disaggregated: Recognising that livelihood profiles differ between group and genders.
4. Long term and flexible.

Sustainable Education Reaching the Unreached

Education does have a bearing on sustainable livelihoods. It can also contribute immensely in achievement of the same. Ayres and Simon (2003) have enumerated that education has the potential to enable people to enhance their capabilities and functioning and can contribute to the achievement of sustainable livelihoods. The evidence for rural Tamil Nadu (India) indicates a complex relationship between education and poverty. The study highlights the importance of the underlying conditions of life, existential realities of different socio-economic groups, contradictions and tensions families face in making decisions about education. Entitlements and livelihood opportunities need to be analysed holistically.

There is a big role that media can play in portraying the positive images of empowerment and achievable community initiatives. Success stories must be aired, enacted and disseminated extensively to bring about a change in the society and mind sets in general.

Education must be of a new type for the sake of the creation of a new world.

> *Knowledge is Liberation...*
> *Knowledge (education) includes all training that is useful for the service of the mankind and (liberation) is freedom from all manner of servitude even in the present life*
>
> —Mahatma Gandhi

It is especially so because :

1. More than half a billion of the people in South Asian sub-region live on less than a dollar a day.
2. More than half of the children of the sub-region are malnourished.
3. Almost half are illiterate (35 per cent men and 59 per cent women).
4. Nearly one-third of the people are unemployed.
5. The sub-region is unable to support its unskilled people in agrarian activities.
6. Rural to urban migration has led to almost half of every city and town turning to slums and shanties.

7. Almost half of the land area in the sub-region is degraded in one form or another.
8. Most parts of the sub-region are highly vulnerable to natural calamities like floods, earthquakes, cyclones, landslides and droughts.
9. With overexploitation, the water resource base in the sub-region is under severe threat and nearly a fourth of the population do not have access to safe drinking water.
10. Reduction in the natural habitat, loss of species and depletion of genetic diversity characterize the plight of bio-diversity in the region.

Source: South Asian Strategy Paper for World Summit on Sustainable Development.

It is as if the above facts are not of the South Aria but of India. And somewhere hidden in them is the crying need for literacy and education. Perhaps because each of these problems are willy-nilly linked to inadequate impact of literacy and education in the country. Readers may notice that literacy and education have been delinked from each other. This is so because they are not the same. Literacy implies the ability of a person to read and write. UNESCO defines a literate person as "the one who has acquired all the essential knowledge and skills which enable him/her to engage in all those activities in which literacy is required for effective functioning in his/her group and community and those attaining in reading, writing and numeracy make it possible to use these skills towards his/her own his/her community's development".

Education can be described as awareness and the ability of the person to perceptively analyse the presented information, against the backdrop of his experience, desires and needs, and to benefit from it.

According to the Gandhian philosophy, education should always be considered as a living process, and the responsibility of sending girls and boys to school should rest on the twin shoulders of parents and the communities. The teaching on the other hand should help promote skills and abilities that are useful throughout life, so that small children can learn skills that will allow them to fully benefit from their subsequent education.

Gandhiji firmly believed that "Educating a girl meant educating several generations", because traditionally women have been the educators of the family and shoulder the grave responsibility of raising human beings with values and attitudes. Therefore, it would be an injustice of great proportion to withhold information, training and access of education from them. This would prevent women from optimally contributing to the development of self-confident, independent, and self-sufficient individuals. He also believed that it is of prime importance to teach people to read and write in their own language, to make them aware of their specificity to be able to value it and defend their culture.

In 2001, the absolute number of illiterates has come down substantially. *Currently two-third of the population, aged seven and above and about 54.2 per cent of females, is literate. The male-female differential in literacy rates has declined to 21.7 per cent as compared to 24.8 per cent in 1991*(NIEPA-Seminar on Progress of Literacy in India: What the Census 2001 Prevails). According to experts, the slower growth of female literacy is low enrolment ratios and high dropout rates especially at the primary stage.

These figures may not, however, be reliable because of conceptual and methodological reasons. And, like in the past cases, where the National Literacy Mission carried out external evaluations, this time also there may exist many discrepancies in the number of literates reported in the census and the actual number of the literate persons. In other words, the literacy figures of the Census 2001 do not present the complete picture and we need to look for factors, and solutions, which impact upon the growth of the literacy in various regions. Let us now look at two most important roadblocks in the path of literacy, viz. availability of teachers and gender disparity, and their possible solutions with the help of examples.

Availability of Teachers

Even though an infrastructure for the schools may be established with the help of the local community, it may not be able to deliver the results for the lack of teachers. Often due to remote locations of the villages and because of lack of any connectivity to the nearest town, teachers are reluctant to work in such schools. Moreover, problems of family children and the societal attitude act as a barrier for the lady teacher to teach in village schools. The teaching becomes irregular in such circumstances and this impacts the student attendance and maintenance of school resources like the building, toilets, blackboard and the other teaching material. Accordingly, a new experiment has been tried in the State of Rajasthan, called the *Shiksha Karmi* Project.

Shiksha Karmi : The Bare-Foot Teacher of Rajasthan

This project was initiated in 1987, primarily to overcome the problem of infrequent teacher presence in the village schools. This project is being implemented with the help of Swedish International Development Corporation. This project proposes to substitute the absent teacher with a local education worker. This person belongs to the local community; he/she often has community support and has undergone intensive in house training. Wider acceptance of these people often makes them very influential in the society. They are thus able to act as a catalyst in further developing the educational infrastructure in the area. Parents often trust them with the education of their children. They, in turn, influence the family decisions regarding family size, money matters and health. As opinion leaders and gatekeepers of information, these people thus become the village's eyes, ears and interpreters. In other words, they acquire influence to direct the development of the village.

According to the EFA 2000 Assessment India Report, "the project aims at universalization and qualitative improvement of primary education in remote and socio-economically backward villages in Rajasthan".

Gender Disparity in Education

Though the concept of universal education had been accepted by citizens of all walks of life, there continues to be a marked difference between the female and the male literacy level. This is primarily due to indifferent social attitude towards the girl child education. Even in this day and age, there are large sections of the society where the education of the girl child is not considered to be of primary importance. Notwithstanding numerous governmental and Non-governmental efforts, the gap between female and male literacy is still very wide and there are more than 10 districts in India where the level of female literacy is under 10 per cent. Although legal framework exists that favours women in education systems, gender parity is difficult to achieve, due to beliefs that limit the level of educational heights that a woman can achieve. Ideological go to religious and cultural factors legitimize the arrangements found by families or societies to meet their basic needs, even though these arrangements may cause the marginalisation of women in the society. Religious education especially in Muslim dominated areas is extremely influential and the community never questions values that have been handed down from generations. In poverty ridden areas of rural India, the need of child labour is so great that girls can never access, remain or succeed. Often lack of any functional family and poverty forces them to abandon school. Another aspect, which can be held directly responsible for low female literacy rates, is the absence of sufficient female teachers in the rural education system. It has been observed that there are less number of female teachers in low literacy areas. These teachers serve as role models for the girl students as they provide them with advice and orientation of their personal life. In situations where the parents are forced to send their daughters to distant schools prefer those schools that have female teachers and especially if the teacher belongs to their community. Unfortunately, they are frequently absent from work because they bear the burden of numerous familial and social obligations, and also due to low wages, precarious work conditions, viz., lack of residential facilities, frequent transfers that entitle shifting of residence, secure teaching environment and absence of separate toilet facilities, and lack of educational materials like black boards, books, etc.

Culture has always played a significant role in the socio-economic conditions of rural areas. It is, therefore, not surprising that it also causes variations in female literacy. These cultural obstacles are directly related to the notion that women should only reproduce and care for their children. It is felt in many communities that since women have to leave home after marriage to live in another family, it is not viable for the parents to dedicate time and resources on their education and thus barriers of religion and culture prevent women from achieving progress outside of the family environment.

The Total Literacy Campaign (TLC), is one step in the direction of involving the total community in the process of illiteracy eradication. It mobilizes, identifies and motivates different groups in villages to embrace literacy programmes, especially for the girls and women.

The Total Literacy Campaign

Implementation of TLC requires nearly one year in an area. In Durg district of Chattisgarh, this period was used for consensus building through discussions and meeting with different officials, leaders and sections of people in different Blocks, clusters and villages. Often 1,100 to 1,200 persons attended these meetings where the main topic of discussion was: "if you want to improve the existing living and working conditions, then literacy is the tool".

The preparatory stage was also used to identify the illiterate population and thanks to thousands of volunteer approx. 317,000 illiterates aged 15-45 were identified. A three-tier training module was thus developed to train master trainers and *akshar sainiks* (volunteer teachers) of which 20 per cent were women (www.literacyexchange.net/india/indiadata.htm). A novel approach was introduced in Durg to motivate the participant. Oaths to deities were taken and literacy pillars were erected in the village to remind people of their commitment. In order to build up confidence in the literacy campaign, issues like health and leprosy care were also undertaken at the initial stages of the literacy classes. There were instances where villages often fined those who did not attend classes. Special groups were also formed to escort women learners for evening classes. They and other learners were also awarded a certificate after completion of the primers, which were written in standard Hindi. The study material was localized for the learners and included words based around their lives. The content was developed around needs and experiences of mostly poor peasants and women. *Topics included work in the fields, use of agricultural technology, and health and hygiene, as well as interpersonal relationships, community, and the environment* (www.literacyexchange.net/india/indiadata.htm). The primary focus of these lessons was always women and so all the material developed was gender sensitive and gender equal. Post literacy classes were followed by basic literacy classes. In Durg, since there were early income generating successes, soybean cultivation was introduced in the post literacy class for selected candidates along with classes in handicraft.

The resultant survey supported by UNESCO clearly indicates that over 90 per cent of the neo-literates were women and belonged to the age group15-35. These women used their newly acquired literacy to write their names and addresses. People and their family members treated them with more respect. These women became more health conscious and gained self-esteem. They were able to take personal initiative and responsibility. And one of the major achievement was an increased awareness of the importance of education, money management and health requirements.

Use of Technology to Increase Literacy

Although, the National Literacy Mission is an extremely well-planned exercise and has produced customized culture-specific teaching materials in all Indian languages, it has a drawback. It takes anywhere between 8 to 18 months to complete one element of the entire course. At this rate the problem of adult literacy would thus take nearly 15 to 20 years to be eradicated. This method of acquiring education takes a very long time and, therefore, it runs the risk of losing students in the programmes midway due to reasons like, earning a livelihood, marriage, social pressures, migration, etc.

Tata Consultancy Services (TCS), India's leading software services recognized this aspect. It, therefore, initiated a pilot project in five villages of Andhra Pradesh, which proposed to impart functional literacy in a matter of months. The aim of this project was to utilize information technology to drastically trim down widespread illiteracy and to develop suitable methodology to teach the rural uneducated. Though its methods and materials were based on the methods and materials of the National Literacy Mission, it differed from this concept. The project was based on the assumption that "if a person can read, he/she can absorb information and knowledge in a written form".

It thus initiated its modules by first building up a vocabulary of about 300-400 frequently used words; extracted from common subjects and situations of that particular community. It then introduced alphabet and script with the help of sounds or phonetics to build these words. The vocabulary was further reinforced with the help of word games. Volunteers in this project used computer based training sessions to teach relationships of the sounds to the written words, separation of the words into composite characters with distinctive sounds and construction of new words with the help of word play. By building up initial vocabulary, the volunteers were able to simplify the logic of forming complex composite characters. The students were thus emboldened to apply native intelligence and cognitive capabilities to grasp the connection between spoken words and their printed forms. They were thus able to intuitively assign a sound to the graphic character in front of them. These inductive experiences motivated the students and created confidence in them to carry on with the academic activity.

Over four weeks time, it was observed, that the students (adults) were able to acquire basic reading ability. This transformed dramatically into reasonable proficiency in 8 to 10 weeks time. The only constraint that the volunteers faced was the lack of adequate number of PC monitors. They were able to increase the number of students by doubling the number of monitors in the class. By running two shifts a day, and catering to 4 alternate batches for a six hour per week instructional lessons, TCS was able to create 60 literates in 10 weeks time.

Under such conditions, it is estimated that 300 adults could be made literate by one instructional location/site in a year. In those remote areas where frequent intervention is not possible, there is a possibility of increased effectiveness if the children are given access to Internet.

Use of Technology to Increase Awareness, i.e., Educate

How would Gandhiji react to the rapid invasion of Information Communication Technology (ICT) in every sphere of Indian polity? Since the main protagonist of the question is no longer in our midst, the question becomes rhetorical. However, the answer to it will still be very interesting. Many among us who knew Gandhiji closely would definitively say that perhaps Gandhiji would equate ICT with the sewing machine, of which he said "one of the few useful things ever invented".

The Information Technology has created numerous opportunities to reach "unreached" masses of the country. India suffers from unequal distribution of scientific and developmental information. This aspect is further compounded by absence of databases of local knowledge and practices, which can combine with ICT's ability to transform generic information to suit local needs. Many prominent Indian social scientists are of the opinion that there is an urgent need to document indigenous knowledge and traditional agricultural practices thereby avoiding the risk of endangering them. Rural families and marginal farmers can use this information and hope to improve productivity of labour and inputs.

Even under such circumstances, ICTs have been at the vanguard of spreading literacy and application of rural development programmes among the vast underprivileged population of the country.

One such project "The Information Village Research Project" was launched in 1998 in Pondicherry, by M.S. Swaminathan Research Foundation, as an experiment in electronic knowledge delivery to the poor, to bring information age in the form of communal telephone and internet access to the rural populace of the region (www.mssrf.org). This project had a clear cut goal that "India's network must reach the villages if it is to make a meaningful contribution to the quality of life", so that a model of information diffusion and exchange in rural areas, using complex information and communication technologies, could be developed.

Half of the people in this region live below the poverty line. They derive information from the local shopkeeper, the market and the agri-input supplier. This information was further exchanged within the family and the neighbours. In other words, the information cycle began and ended within the same area. Though there was a high prevalence of TV sets, there was a considerable information gap because of the absence of area specific development information on the TV (which dealt primarily in entertainment like serials and film based programmes). At the same time, the common perception was that the government channels did not provide the information needed for the upliftment of the area.

The foundation conducted a detailed survey of the region (14 villages, 22,000 people) for identifying poverty and literacy status and the state of telecom infrastructure in the area (www.mssrf.org). Subsequently, Village Information shops were set up. These shops used hybrid 2-way VHF Radio and wired public telephone network to gain integrated voice and data communication capability. Every village centre was connected to this network. Power shortage was overcome by interfacing commercially available digital circuit with solar and grid power. Trained operators from the village itself manned these centres. They also assisted the project staff to generate locally relevant databases. Usually these databases contained information regarding various entitlement schemes of the Government, details about Below Poverty Line families, grain prices in Pondicherry, Input prices of essential agricultural commodities like seeds, fertilizers, etc., directory of general crop insurance schemes, pest management in sugar cane and rice, directory of hospitals and doctors in the region and bus-train time tables. These databases were frequently updated with the help of local sources, web or through other means. These databases were then transformed into local Tamil language in voice/digital audio form.

It was significant that the most frequent users of the information centres were extremely poor assetless farmers and families, of which 18 per cent were women. They derived numerous benefits from the information like availing farm labour insurance, downloading mark sheets of the school examinations, contacting medical practitioners and veterinarians, price information related to grain sales, weather information, and housing schemes. Empowered with the relevant information, farmers have acquired an excellent negotiating edge over unscrupulous middlemen. Daily information about the sea condition from the US Naval Oceanographic Laboratory helped in optimizing fish catch and saving many precious lives of the local fisher-folk (www.mssrf.org).

A significant development was the emerging interest among the youth of the area to transform commercial and educational information into the network content of and for the local area. Rural women operating numerous micro-credit schemes in the area, went a step further and demanded that an account's transaction system, using the data network be established in order to facilitate money transactions and increase transparency in the whole exercise.

Proactive frontline intermediaries equipped with simple transferable skills from the foundation, empowered the rural folk to develop local applications of global concepts. Besides emphasizing the importance of time sensitiveness, these workers assisted the local people in replication of superior technology for use in rural contexts.

Simputer: Bridging Digital Divide

It might not look like a sleek gizmo and will, in all probability, have a black and white screen, but the Simputer is, finally, here to give the personal digital assistants a run for their money.

A Simputer is like a no frills car which is the best for the roads of rural India, while the high-end PDAs are like a preening Merc which needs good roads to function well. Developed indigenously as a low cost alternative to PCs, it is an acronym for Simple, Inexpensive, Multi-lingual People's Computer.

Leading companies like Microsoft and Hewlett Packard have Personal Digital Assistants (PDAs) that cost upwards of Rs. 30,000 while companies like Oregon offer a better price, but all these are more popular with corporates, since the masses find the price bracket just above their reach.

With the Simputer being available in the market, there is hope of bridging the digital divide in India, not only because of its low cost (Rs. 10,000-15,000) but also because a high level of computer literacy is not a must for using this particular hand held device.

Advantages of a Simputer

The first point that the company makes is that the Simputer is an alternative for a PC and not a PDA. "BEL-PicoPeta Simputers and PDAs are not in the same category," says Swami Manohar, CEO PicoPeta, the company which is making the device.

However, others such as Microsoft say that the Simputer is competition for the PDAs available in the market today though it caters to a different class of people.

"The difference between a Simputer and PDAs of other companies, including Microsoft is on account of electronics (Simputer has a black and white display) and fabrication (plastic cover)," says Kartik of Microsoft. A typical user of a PDA would want features like a sophisticated calculator and organiser which increase the cost of electronics in a device, he added.

Designed to be a shared device, it is one of the first gadgets to have an integrated smart card reader and writer. The smart card interface enables sharing of the device amongst several people in the community.

The fact that it can be shared and is affordable makes the Simputer ideal for the Indian masses. It also has a multilingual text-to-speech system, says Manohar.

Moreover, the data stored on the handheld can be transferred to a PC so that memory is not unnecessarily blocked. An icon-driven interface ensures that a high-level of computer literacy is not essential for using the Simputer.

All it takes is the ability to point and touch an icon on screen with a stylus. There is a smart card to ensure that no elaborate identification process ("login"/"password") is necessary.

Based on an open standard operating system the Simputer is targeted at farmers, sales automation, education, banking, e-governance, sales force automation, healthcare, agriculture and rural co-operatives.

There are other portable and pocket PC-like gadgets available in the market but the cost advantage of the Simputer along with its customisation for the Indian market makes it an attractive for the taluk and village-level applications.

Since a Simputer does not run on Windows operating system and has linux based OS, the MS office suite does not run on it.

While the Simputer does not have the hassle of getting Software licences it supports features like playing music and other audio and record speech, an MP3 player and an integrated speaker and soundcard microphone. It also has a notepad, spreadsheet, calculator and image viewer, the company claims.

Who Makes the Simputer?

The Simputer project, which has bought accolades for the country, was conceived by seven trustees drawn from the Faculty of Computer Science and Automation at the Indian Institute of Science and Encore Software Ltd. (formerly Ncroe Technologies).

PicoPeta has the mandate to provide Simputer-based solutions. Bharat Electronics and PicoPeta Simputer have formed an alliance to manufacture and market a new range of Simputers. These devices will be marketed as BEL-PicoPeta Simputers and will cover a spectrum of applications and price points. Currently a few pilot projects using Simputers are running in the rural areas; its retailing strategy will be clear only in June, says Manohar.

Features

Running on the GNU/Linux operating system it has many PC-based applications that are available on Linux. It has 32 MB of DRAM, expandable up to 64 MB and 32MB of Flash memory for permanent storage which is expandable up to 128 MB. The device has text-to-speech capabilities in nearly all languages. It currently supports English, Hindi, Telugu, Kannada and Tamil. The Simputer can also connect with GPS devices. Simputer has multiple modes of connectivity, including infrared for wireless connectivity and USB and serial ports. It has an internal modem with a telephone line for Internet connectivity and a support for WLL (CDMA and corDECT).

Success Story So Far

PicoPeta has executed several projects based on this platform, including e-governance, school education, microfinance, office administration, science and technology.

Besides supplying the hardware and software the company also conceptualised a new model for education, wherein the teacher is empowered to create Simputer-Lessons for use inside and outside the class.

According to initial tests the company found that children understand the subjects better and are excited to be learning their lessons using technology. In the second case, a non-government organisation has been able to track farmer loans across 124 centres spread over 110 villages from barely literate citizens with the help of PicoPeta's microfinance application. Mobile bankers armed with Simputer visit villages, educating, informing and collecting data.

When they are back in office, they use Pico-Peta-developed application to transfer data and observations to a PC. Their managers can then generate various MIS reports, and monitor progress or the loan disbursal programmes. The device even "reads out" text in Indian languages, for the benefit of illiterate farmers.

The company is in the process of adding features like a colour display as a part of the Simputer.

But the fact that it is one of the few devices which hold the promise for bridging the digital divide, is the key to the products success.

Source: Hindustan Times, May 26, 2003.

REFERENCES

Anand-Pioneering the White Revolution (online) (cited on 22nd May, 2003). Available from <URL:http://www.indiaprofile.com/businessandindustry/whiterevolution.htm>.

Ayres, R. and Simon, M.T., 2003, "Education" poverty and sustainable livelihood in Tamilnadu" Inequalities, opportunities and constraints, Review of Political Economy, London, April, Vol. 15, No. 2, pp. 21.

Belavadi, N.V. and M.K. Niyogi, 1999, Chapter Twenty Smallholder Dairy Co-operatives in Falvey L. and Chantalakhana C. (eds.) Smallholder Dairying in the Tropics, International Livestock Research Institute, Nairobi, Kenya, ILRI (International Livestock Research Institute), Nairobi, Kenya (online) (cited on 29th May 2003). Available from <URL:http://www.ssdairy.org/AdditionalRes/Smhdairy/chap20.html>.

India Dairy Industry—A Profile (online), (cited on 29th May 2003). Available from <URL:www.aavinmilk.com/dairyprofile.html>.

India: The Green Revolution (online) (cited on 27th May, 2003). Available from <URL: http://) (www.1upinfo.com/country-guide-study/india/india141.html>.

Schelhaas, H., 1999, Chapter One: The dairy industry in a changing world in Falvey L. and Chantalakhana C. (eds.) *Smallholder Dairying in the Tropics,* International Livestock Research Institute, Nairobi, Kenya, ILRI (International Livestock Research Institute), Nairobi, Kenya (online) (cited on 29th May, 2003). Available from <URL:http://www.ssdairy.org/AdditionalRes/Smhdairy/chap1.html>.

5

Sustainable Health for All

"It is health that is real wealth and not pieces of gold and silver....A diseased person has a prospect of getting well by personal effort. He cannot borrow health from others."

—Mahatma Gandhi

A staunch believer in self-help, Gandhiji believed that, like every other area of human development, health should be one's responsibility, as far as routine everyday problems are concerned. He was of the opinion that a balance should be maintained between one's lifestyle and needs so that good health could keep away medical interventions. Gandhiji was a great advocate of alternative medicine, and placed great emphasis on the body's capacity to heal itself. He often incorporated simple treatment options from ancient Indian healing systems. His favourite was fasting, which he practiced to keep himself fit and to purge his body of minor ailments.

In his opinion, a treatment is holistic if it takes into account not only physical systems but also psychological, social and spiritual aspects of health and illness. Therapies such as these are all about caring and healing because they incorporate body-mind-spirit-medical and caregiver and caretaker's knowledge. This, he felt, mobilized the inner healing resources. An individual thus becomes an active participant in maintaining his/her health and/or curing his malady. At the same time, he understood that the link between human health and economics was very significant. The population remained healthy if the target of sustainable rural development was achieved. Rural welfare thus encompassed a broader vision, which was based on the philosophy of integrating economic empowerment with human productive potential.

Gandhiji was pained to see the miserable condition of health and sanitation in and around villages and towns. The entire community was exposed to health hazards and foul environment due to the use of open spaces as toilets or dry privies. Such conditions gave birth to ill health and premature mortality of people due to chronic communicable/non-communicable diseases as well as diseases like cholera and TB, alongside the omnipresent problems of maternal and infant mortality.

He depreciated the dependence of the people on the "night soil cleaners or scavengers" who physically cleaned and carried human excreta. He was pained at the pitiable social condition of these people. He was aware that the solution did not lie in expensive water-borne sewerage or even septic tanks, but in low cost integrated sanitation and health initiatives arising from the grassroots of the country. It was, therefore, not surprising that he insisted and encouraged people to perform and promote activities related to sanitation, and thus, keep the surroundings clean.

Overview of Health Care Management in India

'India is the second most populous country in the world, accounting for 16 per cent of world's population. More than 300 million of its citizens live in poverty... Although declining, largely preventable diseases such as leprosy, tuberculosis, cataract blindness, and malaria continue to account for a large percentage of reported illness. Pollution in cities causes premature deaths, chronic respiratory diseases and emergence of communicable diseases, due to crowding and lack of access to clean water and sanitation. Government of India reported maternal and infant mortality rates of 570 and 71 deaths per 1,000 births, respectively (TvT Asso.Inc. 1999). India Development Report, 2002 states that life expectancy at birth of 62.4 years has nearly doubled since 1951,...the average infant mortality of 71 per 1,000 live births should be compared with Kerala's 12 per 1,000 live births, (Parikh *et al.*, 2002). As can be seen from the Table 5.1, the high mortality rate is an indicator of poor health systems, inadequate care during pregnancy and child birth and post birth periods.

An emerging health trend in India is a gradual "polarization of health conditions". The poor still suffer and die from preventable infections, pregnancy and child birth related complications and undernutrition. The bulk of disease burden is placed on the poor,

Table 5.1: Some Insights into Indian Demographics

Year	*Life Expectancy at Birth (yrs)*	*Literacy rate (Per cent)*	*Birth rate Per'000*	*Death rate Per '000*	*Infant mortality Rate per '000*
1951	32.1	18.3	39.9	27.4	146.0
1997	62.4a	62.0b	27.2	8.9	71.0
2001	63.7	65.38	25.0*	8.1*	68.0*

a refers to 1996, *b* provisional.

Source: *Economic Survey, 1999-2000*, Table 10.3, p. 167, *Human Development Report*, 2004.

* Year 2002.

on women and on scheduled castes and tribes. Overall, the poorest 20 per cent of Indians have more than twice the rates of mortality, malnutrition and fertility of the richest 20 per cent. The effectiveness of public programmes has been limited in part because overall public investment in health remains low, and because constraints in the structure of the health system have made it inefficient. The public remains badly informed about much of the health system. It knows little about whether health services are appropriate, who is benefiting from them, whether quality is sufficient, or whether people are getting good value from public and private spending on health (Peter *et al.*, 2001). Despite the establishment of a large public network of health providers, public spending on health is relatively low, and has stagnated at levels of around 1 per cent of GDP. The bulk of public spending on primary health care has been spread out too thinly for these services to be effective, while the referral linkages to secondary care have also, suffered (Tulasidhar, 1996: Mukhopadhyay, 1997).

Stagnant public spending on health (less than 1 per cent of GDP) places India among the bottom 20 per cent of countries. The bulk of public spending on primary health care has been spread too thinly to be fully effective, while the referral linkages to secondary care have also suffered. Between 75–90 per cent of public health spending is done by the States whose funds are largely go to pay salaries, leaving little for essential drugs, supplies, operations, and maintenance. Curative public services also mostly favour the rich (Figure 5.1). About Rs. 3 is spent on the richest quintile for every Rs. 1 spent on the poorest 20 per cent. There are also differences in distribution for different types of services; immunizations and outpatient services at primary health centres are pro-poor, but in-patient services are less likely to reach the poor (Peter *et al.*, 2001).

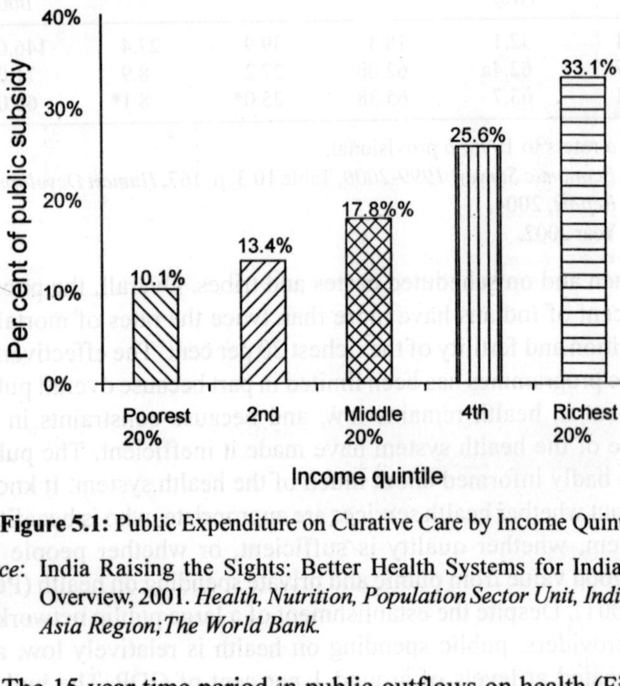

Figure 5.1: Public Expenditure on Curative Care by Income Quintile

Source: India Raising the Sights: Better Health Systems for India's Poor Overview, 2001. *Health, Nutrition, Population Sector Unit, India, South Asia Region; The World Bank.*

The 15-year time period in public outflows on health (Fig. 5.2) highlights some remarkable details. The per capita real spending initially exhibits a positive trend, which steadily increased in the first 11 years. On current prices, total health spending by the public sector in 1998-99 was Rs. 161 billion or the equivalent of Rs. 165 per capita ($ 3.9 or Rs. 97) in 1993-94 prices. Starting from a high of 1.1 per cent of GDP in 1985-86, the commitment steadily drops to a low of 0.78 per cent in 1996-97 then somewhat recovers in the last three years to 0.9 per cent (Fig. 5.2). But even at 1.1 per cent of GDP, India ranks among the lowest in the world.

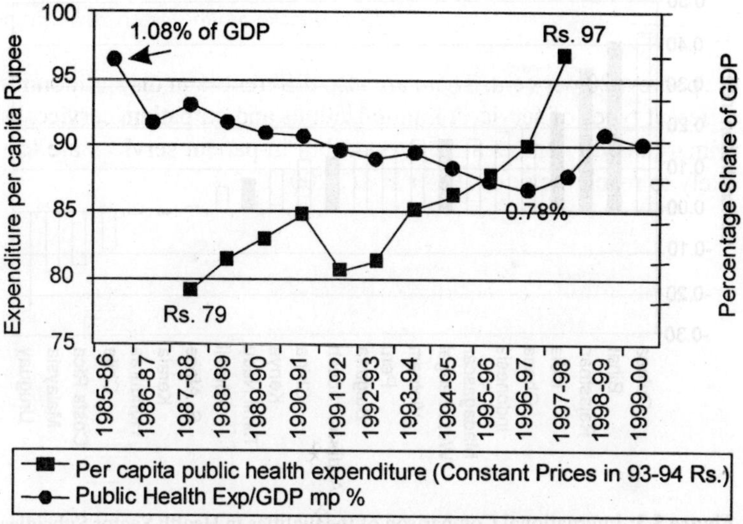

Figure 5.2: Public Spending on Health

Source: Central Statistical Office, Ministry of Finance; Reserve Bank of India.

In comparison, internationally, 'equity of public spending on health in developing countries, indicates that :

1. Public spending is most pro-poor at basic levels of social services.
2. Discrimination by race, gender, caste, and minority status plays a role in shifting utilization of publicly provided and financed services.
3. Some evidence supports the findings that public spending in the health sector appears to be more pro-poor in urban settings than in rural settings.
4. Socialist countries have a better pro-poor record than non-socialist countries. These four findings are consistent with the new utilization and benefit incidence results from India.

Fig. 5.3 compares some of the benefit incidence results from India with findings from other countries. The black bar represents the concentration index for India while the grey bars represent selected Indian States (World Bank, 2001).

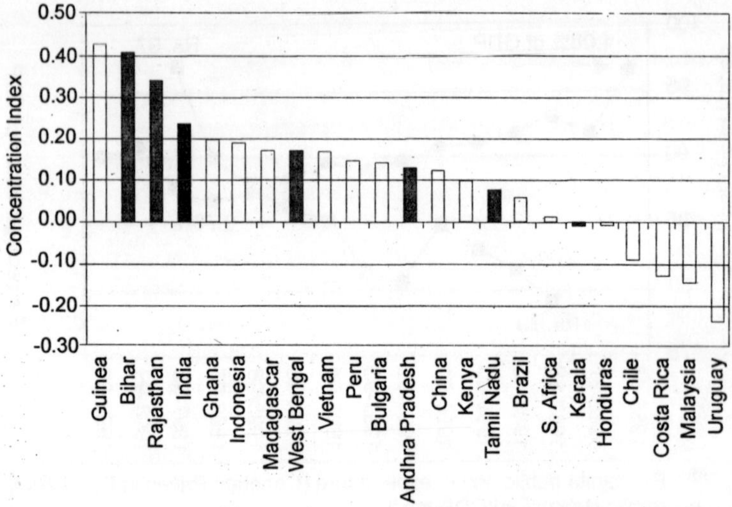

Figure 5.3: International Comparison of Inequalities in Health Sector Subsidies

Source: India Raising the Sights: Better Health Systems for India's Poor Overview, 2001. Health, Nutrition, Population Sector Unit, India, South Asia Region; The World Bank.

As can be observed from the above graph, the bulk of the findings point to public spending patterns in health that are not very pro-poor. Even internationally, preventive and promotive health services take a back seat to curative care. Yet, preventive care is almost exclusively provided through the public sector: about 90 per cent of immunizations and 60 per cent of antenatal care is estimated to be provided through the public sector (International Institute of Population Sciences, 1999; Background Paper 18). The public sector has been organized largely to finance and deliver curative care, as well as implement a number of centrally sponsored family welfare and disease control programmes. These programmes are almost exclusively delivered through public institutions. As a result, India has amassed an enormous but under-funded public delivery infrastructure and staff, that in 1999 included about 137,000 sub-centres, 28,000 dispensaries, 23,000 primary health centres (PHCs), 3,500 urban family welfare facilities, 3,000 community health centres (CHCs) and an additional 12,000 secondary and tertiary hospitals (CHCs are 30 bed secondary hospitals). In rural areas, public sector manpower included 29,000 doctors, 18,000 nurse-midwives, 134,000 auxiliary nurse-midwives (ANMs), 73,000 male multipurpose workers, 21,000 pharmacists, and another 60,000 paramedical staff, in addition to non-technical staff (Ministry of Health & Family Welfare, 2000).

India's Share of World Health Problems

The World Health Organization ranks India 134th of 191 countries in disability-adjusted life expectancy at birth (WHO, 2000). Fortunately, according to WHO, 2000, India is not yet suffering from the low level of disability-adjusted life expectancy seen across Africa (including middle-income countries like South Africa), where the HIV epidemic has further advanced. However, according to Wang *et al.*, 1999, India performed worse than predicted by its income and education levels in the areas of mortality under the age of 5 years, female adult mortality, and total fertility. India performed better than expected for male adult mortality.

Table 5.4: India's Share of the World's Health Problems (per cent)

Population	17
People living in poverty (< US $ 1/day)	36
Total deaths	17
Child deaths	23
Material deaths	20
Disability adjusted life years lost	20
Childhood Vaccine Preventible deaths	26
Persons with HIV	14
Tuberculosis cases	30
Leprosy cases	68

Source: World Development Indicators (2000): World Health Organisation (2000).

Of the various factors which promote high mortality in Indian population is the dismal hygiene situation in the country. In the absence of a holistic approach, numerous isolated and sporadic cleanliness drives failed to make an impact on a health scenario, which kept on degenerating by epidemic proportions.

Nearly 30 years after the death of Mahatma Gandhi the first Pan India movement to restore the Human Rights of these scavengers, promote environmental sanitation, health and hygiene, non-conventional sources of energy, waste management and social reforms through education, training and awareness campaign, was launched by Dr. Bindeshwar Pathak in 1970.

Table 5.5: Comparison of India and Other Countries on Selected Indicators

Country	*Population (millions) 1999*	*GNP per capita US$ 1999*	*Population below $1a day (per cent) 1992-98*	*Adult female illiteracy rate (per cent) 1998*	*Disability adjusted life expectancy 1999*	*Infant mortality (per 100%)1996*	*Total fertility rate 1998*	*Prevalence of child mal-nutrition under age 5 (per cent) 1992-98*
India	996	450	44.2	57	53.2	72	3.3	63
Low-income countries	2,417	410	36.0	49		66	3.1	36
Nigeria	124	310	70.2	48	38.3	76	6.3	39
Bangladesh	128	370	29.1	71	49.9	73	3.1	56
Pakistan	135	470	31.0	71	55.9	91	4.9	38
Indonesia	207	580	15.2	20	59.7	43	2.7	34
Middle-income countries	2,667	2,000	8.2	20		31	2.5	12
China	1,250	780	18.5	25	62.3	31	1.9	16
Sri Lanka	19	820	6.6	12	62.9	16	2.1	38
South Africa	42	3,160	11.5	16	39.9	51	2.8	9
Brazil	168	4,420	5.1	16	69.1	33	2.3	6
High-income countries	891	27,730	0.1	–		6	1.7	–
United Kingdom	59	22,640	–	–	71.7	6	1.7	–
United States	273	30,600	–	–	70.0	7	2.0	1
World	5,976	4,890	24.0	32	–	64	2.7	30

The Case of Sulabh International

After scrutinizing numerous models of toilets and analyzing several studies in India and outside, Dr. Pathak developed "The Sulabh Shauchalaya Technology". Apart from being socially acceptable, this technology was also low cost and easy to construct with locally available material. The design could be modified to suit the needs and the paying capacity of the users. Besides, it required only two litres of water to operate. It could also function in the regions where there was a paucity of water. This technology did away with the services of the scavengers, and helped in keeping the environment clean. The spin offs generated from this technology are:

Fertilizers for the fields becomes easily available on the spot, bio-gas for cooking is generated free of cost, and this technology provides a workable solution of the age old problem of environmental degradation, through pollution of ground water, soil, and air, at no extra cost to the public exchequer; besides providing the community with an additional source of income and a solution to their need for toilet complexes.

Environmentally, "*the impact of safe disposal of human excreta on health has been significant in reducing mortality and morbidity from gastro enteritic diseases and helminthic infections*". (Sulabh International online).

Sulabh has initiated a radical approach to give a fresh impetus to the works of Gandhiji. By seriously undertaking the rehabilitation work of the scavengers and their families into the society, Sulabh has sensitively handled the cruel social dichotomy. Another aspect of Gandhian social revolution was the upliftment of the Harijans to the same social position as that of the higher castes in the country. Dr. Pathak's question to this aspect of the Gandhian thought was; "*The question is, if scavengers were liberated and rehabilitated, will they attain the same status in the society as that the upper castes get?*" in reply to his own question he then stated — "*We are persuading Brahmins to help them in learning prayers and entering temples and to dine with them*". Before long he led 100 girls from scavenger families into Nathwara, a temple in the State of Rajasthan where traditionally the scavengers have been prohibited from entering and worshipping the deity. Not content with revolutionizing the ancient ways of the society, Sulabh has also chosen the route of training and rehabilitation of scavengers and their wards.

Training and Rehabilitation

Scavengers belong to that strata of the society, which is not skilled for any alternative occupation, and their literacy levels are low. Whatever little expertise they possess is often not sufficient to provide sustainable income for the entire year. The problem is enormous and often the solutions are varied. It is under such circumstance that a Vocational Training Centre was set up in Delhi. "The training centre aspires to provide skill development training to boys and girls of weaker sections, so that their talent and potential, to the fullest extent, could be exploited to achieve better economic and social adjustments."

It is here that Sulabh arranges a training curriculum for different vocations such as computers, cane work, tailoring, driving, mason work, carpentry, automobile mechanic work, etc. The Sulabh International Centre for Action Sociology has initiated a welfare programme for the under privileged children of the slums, who are constantly exposed to physical and social hazards due to the nature of their habitat. These boys and girls are taught the importance of health and hygiene. They also undergo regular check-ups for cleanliness. Supervised physical training, drills and games, instructions in functional literacy and, lastly, cultural sensitization by introducing dance and drama in routine has considerably improved the rate of their physical, moral and social rehabilitation.

Currently, Sulabh International is working in close coordination with the State and Central governments and many international agencies like UNICEF, WHO, UNDP/World Bank in order to hasten the adoption of the low cost technological option in India and the other developing countries.

The Role of other Agencies in Promoting Rural Health

Not every agency working in the rural health sector has a pan Indian vision as Sulabh. However, each one of them agrees on the concept that all the health programmes initiated across the country run on 3-Is, i.e., Information, Intervention and Innovation.

Numerous governmental and non-governmental organizations have, therefore, focused on either of these areas. Let us begin to examine them.

First 'I' or Information

The reason why the majority of the population is bereft of the benefits of modern medicine is the lack of information and awareness of the causes and symptoms of various afflictions. India is home of over a billion people. It is also the region where even today over 50 per cent of deaths are caused by infectious diseases, most of which are preventible and treatable. Nearly 53 per cent of children under the age of 5 suffer from malnutrition, even when the country is not short of food. These aspects are indicative of the need for information and better dissemination of the same.

Technology has always played an integral role in the spread of information. Organizations using it is vast, we have, therefore, chosen to discuss in two of them representing two different regions of the country and having people belonging to two extreme ends of the society striving to activate the same goal. These organizations, namely, Nalamdana in Tamil Nadu and Video SEWA in Gujarat, have used communication technology as a tool for spread of awareness.

The Case of Nalamdana

In 1993, Uttara Bharath, an alumnus of John Hopkins University, along with a few like-minded people, and a common interest in theatre formed Nalamdana in the State of Tamil Nadu in India. Initial funding for the organization was from Echoing Green Foundation, New York. At present Nalamdana is being funded by the Ford Foundation, Action Aid, UNICEF, etc. An Experimental performance of street plays set them on the path of using entertainment for communicating information about sensitive socio-emotive health issues like HIV/AIDS. Nalamdana's mission is to provide "health education and promote preventive behaviours through entertaining methods: street theatre, discussion groups, and songs with messages". It has used innovative communication methods to deliver key health messages like the need for cleanliness, health and hygiene, HIV/AIDS, STD, suicide prevention, and cancer to semi-literate audiences in inner city slums and villages in Tamil Nadu. Its primary goal being the empowerment of people to make informed decisions. Till date, Nalamdana has performed in more than 111 slums covering the audience of more than 200,000 people. Nearly, a million people have watched and benefited from its performance. At present Nalamdana is focusing on promoting preventive health behaviour in two key areas, viz., HIV/AIDS and Pre-Natal and early childhood malnutrition.

How Does Nalamdana Do It?

Nalamdana has developed a distinctive style of street theatre based on the manner of Tamil films. It trains college students and recruits many of its actors and actresses from the slum communities, often during street plays local people are called upon to display their flair of acting on the improvised stage. By doing so, Nalamdana brings in the community nearer to its efforts and this permits them to present the *message in a style and manner that is specific to their micro-culture. It also gives Nalamdana the opportunity to develop its message strategies with a closer understanding of the more subtle beliefs and issues of the communities.* Street theatre plays can be adapted to suit any particular community's needs. There is an immediate feedback, it is entertaining and non-intrusive and targets everyone without being particularly focused on one individual in the audience. The organization conducts random interviews after the performance is over and provides counselling by actors who have specially been trained to impart information regarding HIV/AIDS. It also conducts workshops for illiterate men and women regarding STDs and HIV. At present Nalamdana has begun to use audiotapes and video films, as a tool to reinforce the information about HIV/AID awareness among its target populace. Nalamdana has encouraged people from the community to write the lyrics of the songs in the audiotapes, which were then performed either by the writer himself or by leading artistes of the State. Creation of these audiotapes and video films is a part of Nalamdana's ongoing Information, Education and Communication material development programmes wherein they produce marketable tapes for creating mass awareness.

Video SEWA

The Self-Employed Women's Association (SEWA) was set up under the auspices of Textile Labour Association (TLA) in 1972. However, since the date of its inception it has followed three movements: the labour movement, the cooperative movement and the women's movement. The inspiration behind SEWA is the Gandhian philosophy which insists that the poor empower themselves with the help of twin strategies of struggle and development. The struggle is against many societal limitations, implementation of laws, policies and official apathy. While the development aspect encourages the cooperatives, self-reliance, and social securities of which health care and health awareness forms an important facet.

An offshoot of the Self Employed Women's Association, the Video SEWA was born in 1984, during a three week workshop held by Late Martha Stuart, an International Video Communications Consultant from New York. Present at that workshop were 20 illiterate women, who soon after began to make video films. In the absence of any editing facility, these women shot films in sequence. It was only after three years, in 1987, when they could first afford editing equipment. These women could conceptualize script, record and edit these films, however, they could not find these tapes on the shelves of their officers because they were totally illiterate and hence could not read the written word. Yet, they had managed to master the complex art of making films by memorizing each and every step and button on the machines. A new training session was held in 1994 where 15 grassroot women were taught to make films and handle the video equipment.

Besides making films on issues that concerned their livelihood and working conditions, they also made films concerning primary health and hygiene awareness. Films that spread awareness about everyday ailments; like dehydration among children; and necessary precautions to avoid them; were especially well received by the audiences as they told them about what steps to take in case of eventualities. Video SEWA also helped in spread of awareness about the use of smokeless stoves in place of the Kerosene stoves, which caused many lung problems among rural women. SEWA has made watching the videos an important component of awareness campaign, because they give members and rural women an opportunity to understand the issues pertaining to their own health. These videos also helped in bridging caste and cultural barriers and made people empathize with each other's concerns. There have been times when videos have helped women understand complex issues regarding their health because face-to-face communication was not possible due to the sensitive nature of the topic and the relative anonymity of watching them eased a lot of hesitation. Especially in rural and urban slums, video productions act as a catalyst for initiating discussions on various issues portrayed. This grassroot technological initiative has transformed into an effective tool enabling awareness building, participation and empowerment.

Video films have thus become a very powerful tool in the hands of powerless. Nearly 19 years after the first film was made the older generation has begun to give way to the new generation. It is interesting to note that the son and daughter of Martha Stuart come once a year to technically upgrade these women on the latest innovations and development in the field of film making.

Second "I" or Intervention

A constantly growing population, and a regular migration from the rural areas towards the urban centres, combined with the rapidly increasing divide between the rich and the poor has stretched the capacity of health systems in India. Non-functional at the best of times, this system is collapsing even as the efforts to shore it up are going on at a very rapid scale. According to the World Bank Report on Better Health Systems for India's Poor (2001), "India is in the midst of a 'health transition'—at varying rates depending on the State and population group". The report has also termed the transition as demographic citing the decline in mortality and fertility rates along with the ageing of population. It further states this transition as both epidemiological and social because of the changing pattern of illness from malnutrition and communicable diseases based on chronic diseases of adulthood in a population that has rising expectations and capabilities. Yet, there remains an "unfinished agenda" wherein a large percentage of the population continues to die from preventible infections, pregnancy and childbirth related complications and under-nutrition.

Not surprisingly, given the state of affairs, the recent World Bank Survey (2001) has recommended public and private participation and intervention, so that the health system can be made more pro-poor, gender sensitive, client friendly, accountable, affordable, accessible and equitable. A comparison between the public and private health initiatives have thrown up startling facts, (a) The private health sector caters more to the rich; and (b) about 79 per cent of the outpatient care for the poor are still provided by the low quality and untrained practitioners.

A large number of Government and Non-Government Organizations are working in conjunction with numerous national and international organizations to fill up the lacunae that exists in area of quality health management at the grassroot level.

We have taken up the case of two organizations that are providing quality interventions in health sectors; by utilizing two different approaches, viz., top down and integrated, in rural areas of the country—The Central Himalayan Rural Action Group (CHIRAG) in the Nainital district of Uttaranchal and the Rural Health Project of Rishi Valley School near Madanapalle in Andhra Pradesh. Both these institutions are situated in the foothills of North India and the South India respectively. However, each has its unique set of health problems, originating from disappearing natural ecology and medical neglect.

The Central Himalayan Rural Action Group (CHIRAG)

Uttaranchal, ecologically one of the most sensitive regions in northern part of the country, is also responsible for maintaining the hydrological stability of the entire northern part of India. Yet, today, it faces a severe stress due to devastating degradation. Its socio-economic development is directly dependent on its natural resources. Hence, the environmental decay has contributed substantially to the rapid decline in the economic and health status of the region.

A major contributor to the economic development of the region are women, who take care of numerous household and financial chores in the absence of the men folk who are usually employed in major urban centres and are, therefore, away from their villages for a long time. Relentless destruction of land has meant decreasing fertility, disappearance of the natural herbal cover, increase in pollution and more backbreaking hard work for women who are responsible for looking after cattle, working on the fields and fetching firewood, fodder and water. Their efforts to earn a subsistence level existence are far greater than that of men. It is the women and their infants who suffer the most due to the degradation of land and lack of healthcare. In such a situation it was, therefore, pertinent that healthcare should be made the leading component of any socio-economic and ecological developmental initiative.

CHIRAG, a non-profitable organization was already active on a small scale in this area since 1988. It had an in-depth knowledge of the area and had already identified health as the engine for economic growth of the region, while integrated intervention was to be its driver. CHIRAG along Canada India Village Aid (CIVA) initiated a Community Health Project in 1988. The purpose of this $ 140,000 project was to improve the health and nutritional condition of 14,000 villagers in the Nainital district of Uttarakhand. It specially focused on the women in childbearing age and children under 5 years of age, because it is they who are potentially the most effective community based agents of change. It was noticed that they were found to be more receptive of any technique or work methodology that would reduce their burden and make their lives more bearable.

This project targeted local women, chosen for their articulatory skills and overall intelligence, to form a dedicated cadre of community health workers and trained midwives, so that the intervention programmes may respond to special needs of the target populations. Such empowerment encourages contact between the traditional and modern medical systems. The purpose of this exercise is that, familiar people and common language become instrumental in spreading effective societal urgency on the implementation of hygiene, sanitation, immunization, nutrition and low cost herbal home remedies for minor ailments. Many modern medical practices are thus demystified, as are the costs of health care.

The trained groups work under the guidance of medical personnel at the Rural Health Centre. Although modest, these facilities, together with the diagnostic laboratory and a drug store, provide competent medical care services, to the people of the area. Health workers visit households and identify children who need immunization cover, they also advice the women of the house about how to deal with minor cases of dehydration. Besides, monitoring the health of women, children and the other members of the village, they educate the populace to look for signs of any major health problem that might crop up. They also facilitate the interaction between the rural health centre and the local people. Such is the degree of involvement that many families are now keen to become an active participant in these rural health drives. Not only is the social standing of the women health workers increased, it has also opened up another avenue of employment generation for the women nearer to their homes. These change agents besides impacting the health status of the region have also become instrumental in many socio-economic efforts, viz., social forestry, wasteland development, and various sustainable energy programmes, that have been initiated by CHIRAG and CIVA.

This project has been identified, by CHIRAG and CIVA, as the most suitable way to gain the trust and confidence of the communities so that with their cooperation the practices, which threaten to destroy the ecosystems of the Central Himalayan region, can be reversed.

Numerous health related programmes are, however, not run free of costs. All the previous Government interventions were perceived by the people as charity and hence had a very dehumanizing effect on the population. CHIRAG from the onset insisted that people with paying power should pay for all the medical treatments; however, those who were not in any state to pay for them would have to contribute in some form so that the dignity of the patient and the treatment could be maintained. This approach has given a very professional slant to the entire exercise and has imbibed a sense of ownership into the people about the project.

At present, this movement and its various offshoots, have reclaimed vast tracts of denuded forests, decreased various episodes of preventible diseases, brought water very near to the habitats and most important of all inculcated in the people, that development of their region is not dependent on the Governmental dole outs and instead lies in their hands.

The Rishi Valley Rural Health Centre

In 1978, when Dr. Kartik passed out from the Rishi Valley School (The Krishnamurti Foundation, India), he perhaps had no idea that one day he would come back to initiate a health intervention programme in the form of a Rural Health Centre. Initially the effort was twice weekly for the treatment of outpatients in and around the rural areas of the school. Health care in Andhra Pradesh is generally steeped in a lot of myths and black magic. People visit local religious practitioners for their ailments, who use ancient methods like swiping the patient with twigs and branches, conducting numerous fire ceremonies to drive away the evil spirits from the body of the sufferers or by branding the patient with red-hot iron. By the time that the patient seeks proper medical intervention, it is already too late to save him/her. Consequently, treatable diseases like dehydration, measles, faulty eyesight, TB, etc., keep afflicting patients unhindered.

In July 1999, Dr. Kartik set up a Rural Health Centre in the Rural Education Centre (REC) of Rishi Valley School. Initially he conducted an Out Patient Department, twice a week, for patients from local villages and students from the REC. He also visited the surrounding villages twice a week. The guiding principle of this centre was to provide:

1. Quality, ethical, humane medical care, at the most affordable rates, with minimal possible medication, to the patients,
2. Attention is given to each patient, in the form of advice on preventive measures (diet, nutrition, hygiene) as the case may be in addition to medicines,
3. *A follow up is done on all chronically ill patients (TB, Hypertension, Diabetes, etc.) by the network of REC teachers.* (www.ruralhealthrv.org.)

It was not before long that the first male health worker was trained on the job. Interestingly, the initial health workers were the men of surrounding areas, primarily because it was they who wielded the decision making power in the household and, therefore, the constructive change towards the health issues had to be initiated with their help. Secondly, it was also practical to begin training with the male members of the society. Before long, the first lady health worker joined the ranks. She was (is) not only capable of diagnosing and treating simple ailments but was (is) also knowledgeable about when an intervention from doctor is required. She has inspired other women to become a part of the team of health workers. The teachers and students of Rishi Valley School are also trained to impart basic education on various public, social and community health problems like oral rehydration, personal hygiene, nutrition, maternal and child health care, etc., during their formal and informal interactions with the community during adult literacy classes and various village committee meetings.

The approach of the Rural Health centre is particularly focused towards the eye related diseases caused because of malnutrition and long working hours in the fields where the population is exposed to harsh sunlight and dust. Poor ophthalmic hygiene compounded by poverty and malnutrition, a common condition, is widely prevalent in the area and has made the local people susceptible to eye ailments like trachoma, cataract and various other visual disabilities.

The centre screens patients for eye diseases and then sends them to Shankara Netralaya (a leading eye care centre in the Southern India) for operative and post operative care. It has also tied up with local hospitals for the services of the doctors and inpatient treatments for numerous other diseases.

Though the primary focus is on eye-care, the centre has now included women health care as its next area of intervention. Common problems among them are caused due to nutritional deficiencies, personal hygiene, and unhygienic pre- and post-natal care. The centre has begun to train the traditional birth attendants in the correct methodology of delivering babies in a clean and germ-free manner. They are also educated about the immunization programme for the babies, so that they can convince the families to let their newborns undergo the complete immunization protocol.

There are a large number of villagers with chest ailments, especially Asthma and TB primarily caused by the smoke of fire wood, used for cooking in women and by smoking bidis and cigarettes in men, regular chest clinic and monitoring of the health workers have ensured that patients complete their treatments.

Since the centre subsidizes the treatment to its patients, it is largely dependent upon the contributions made by the old students of Rishi Valley Public School. Medication is free for the poor and the needy, but the rich have to pay in full for their prescription. The Centre has also tied up with LOCOST, an NGO in Guajrat, which manufactures quality medicines as per the WHO essential list for Developing countries and other like-minded organizations for the supply of medicines for the centre.

At present, the efforts of the centre are showing results in the form of changing mentality towards personal health and health education, follow ups and a growing interest in the socio-economic developments which were lagging due to health constraints.

Third 'I' or Innovation

It was Mahatma Gandhi's dream that India should have an economy based on handicrafts instead of industry and self-reliance rather than international trade. At the first sight, the Jaipur foot seems just a hand crafted foot for self-reliance. But before we go into the details of this innovation, let us try and understand its need in the Indian context.

According to National Sample Survey Organization (NSSO) Government of India (2004) around 5.5 million people have some kind of disability with legs (See Figure 5.6). Approximately 1 million people have either a hand or a leg. Around 4 million people have become disabled due to polio. Every year around 25,000 people become disabled due to accidents. For every 1 lakh people in the country, 1,755 were affected by some disability with 2,000 males among a lakh people being affected compared to 1,493 women. Some 18.53 million are estimated to suffer from disability in the country and more than 10 per cent of the disabled population suffered from more than one type of disability with more affected people in the rural areas than urban parts of the country, the NSSO report states.

While another survey says that 5-6 per cent of India's population is disabled, which adds up to over 70 million Indians (NCPEDP online, Ray, 2001).

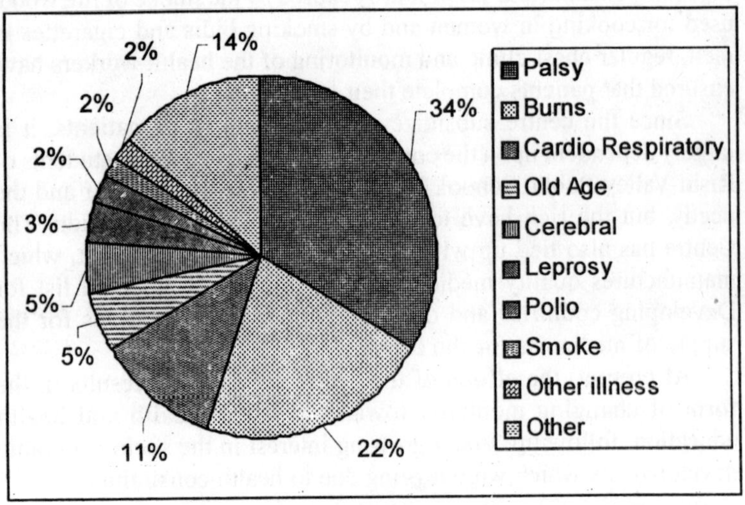

Figure 5.4 : Causes of Locomotor Disability* 1991 Sample Survey

Source : (i) JICA Country Profile on Disability, The Republic of India, 2002, original; and (ii) SARVEKSHANA, 47th Round NSSO (Per 1,000 distribution).

* Locomotor Disability means disability of the bones, joints or muscles leading to substantial restriction of the movement of the limbs or any form of cerebral palsy.

Given such vast disparities in the numbers, it is not surprising that disabled remain in prisons of dependencies, the government remains indifferent and the public ignorant, while the disabilities sector remains a mute spectator doing little to change the system. While specific policies and reforms are still needed, the Bill on The Person's Disabilities (Equal Opportunities, Protection of Rights and Full Participation) Act of 1995 is a definite step in addressing the needs of the disabled members of the society.

A majority of these people belong to the poor sections of the polity. Making these people self-reliant is important to integrate them into the mainstream of the society. Although numerous aids and appliances have existed for the handicapped sections for over thirty years, acquiring an artificial limb was never easy for the poor, due to the high costs or functional limitations. At the grassroot level, it becomes all the more important that a suitable design of the limb be developed by understanding the socio-economic needs of the user.

The Jaipur Foot

According to Hamish McCrae (*The Independent*, 30th March 1995)—when the markets cannot find the right solution, it is best to listen to what customers want. The enduring message of the Jaipur Foot is, listen to the users.

The chain of events that culminated in the world's only $ 28 artificial foot began when an Orthopaedic surgeon began listening to his patients over 30 years ago. Dr. P.K. Sethi was the head of the department of orthopaedics in a hospital; where he first noticed that his patients reverted to their crutches even after being fitted with an artificial limb. This was a setback to him as he had just established an updated factory to manufacture artificial limbs based on western designs. On asking, he was told that western prosthetics work well if worn with shoes and walked on footpaths; since many Indians go about their jobs barefoot and use their feet in different ways, these limbs are not effective in Indian conditions. In words of Dr. Sethi—"We don't have the chair-sitting culture, like other warm countries, we have a floor-sitting culture. Or we squat. We take off our shoes inside our homes and shrines and mosques. The western foot-piece did not have the mobility for sitting that way, or for walking on uneven ground." What Dr. Sethi implied, that sophisticated plastics, foams and other materials could not stand the wetness of rice fields, where they are usually up to their ankles in water/mud, and were meant to be covered by shoe and once shorn off their protective covering looked very grotesque. The amputees thus needed to take off their artificial limbs many times a day for various purposes, which cause great inconvenience.

Dr. Sethi along with Ram Chandra Sharma (Jaipur Foot online), then developed an artificial limb very unlike its western counterpart. Firstly, it looked like a foot instead of a piece of machinery, it was encased in readily available vulcanized waterproof rubber, the single wooden heel of the western model was replaced by two wooden heels joined together by a large rubber block. Later on, and after much experimenting, Dr. Sethi along with his co-developer Ram Chandra Sharma replaced the front wooden block of the limb with hard rubber to enable the user to climb hills and trees. This design was better suited to the needs of the developing countries. The only innovation, that could further improve this perfect design was, the one which reduces the weight of the limb. To quote Dr. Sethi: "What we need is a technology which is more scientific, less expensive, and therefore, more equitable, more harmonious……, and which augments autonomous coping……rather than passive and expensive caring available in our institutional systems". For long, the patients and the doctors had to make do with the available material before Indian Space Research Organization transferred its polyurethane technology to a Jaipur based social organization. This technology is a spin-off, borne out of research and development of various polymeric materials. This technology helps to produce more durable, lighter and comfortable artificial foot. It substitutes the rubber and wooden ankle block with more durable polyurethane micro-cellular foams, thus making the foot comfortable, lighter, more durable and abrasion resistant.

Jaipur Foot also has a unique advantage. Unlike other limbs, it can be customized to suit the wearers requirements, the case in point is that of the famous TV actress and Dancer Sudha Chandran (Lakshman, 1996), who had lost a limb in an accident. She soon got back to the dance floor with an artificial—"Jaipur Foot", which was customized specially for her by Dr. Sethi; given her need to dance again. Jaipur Foot is today India's Humanitarian export especially to the countries which have been a victim of wars and landmines like Afghanistan, Cambodia, Sri Lanka, Angola, etc.

Jaipur Foot has been a conscious attempt to create a low cost technology to suit the needs of amputees in developing countries. It was also an attempt to harness the knowledge of rural communities to make radical differences in peoples' lives. The wealth of experience points towards an important but practical aspect, that the most suitable technology is frequently not something dreamt up in hi-tech west but most often developed in the backyard of the developing countries.

Conclusion

No single solution fits all services in all the countries. The difference in various models of implementation may be on account of accountability. The root cause of many problems is poverty. Illiteracy, lack of hygiene, unemployment all lead to substandard living conditions leading to higher health problems. Gandhiji's idea was always maintenance of health. A healthy mind exists in a healthy body. Sanitation in keeping environment clean are some basic parameters. Health care interventions alone will not yield meaningful results in isolation. Simultaneously a number of issues will have to be addressed in order to raise the general health of people. According to WHO (WHO and UNICEF, 2000), the improved water supply and adequate sanitation will result in:

1. 25 to 33 per cent reduction in diarrhoeal diseases in the developing world which accounts for 4 billion cases;
2. Decreased incidence of intestinal worm infestation (estimated at 10 per cent of population in developing countries) leading to malnutrition, anaemia and retarded growth; and
3. Controlling blindness due to trachoma and schistosomiasis which are water related.

In India, poor hygiene and sanitation account for 9 per cent of all deaths and an estimated 27,463,000 years of life lost each year (Murray, 1996). About 73 million workdays are lost due to water borne diseases. There is substantial evidence in literature which relates that adequate water supply and sanitation as being important determinants of health, education and social well-being.

Poor sanitation facilities, and lack of availability of clean water have their own causes as well. Historically water and sanitation have received a very small proportion of the budget (Mavalankar and Shankar, 2004). The budgetary allocation from year 1980 to 2000 indicates that the water and sanitation have been getting only 2 per cent of allocation of the Central Government budget. Some of the reasons for this state of affairs are :

1. Lack of political will,
2. Low prestige and recognition of sanitation,
3. Poor policy framework at all levels,
4. Poor institutional framework,
5. Inadequate and poorly used resources,
6. Inappropriate approaches,
7. Neglect of consumer preferences and low public awareness, and
8. Lack of public health leadership.

Also depending on the geographical area, the medical factors may vary greatly.

The interventions discussed in the chapter indicate that some indigenous interventions do help. The disease patterns vary and the socio-demographic profiles also vary. So solutions imported in different socio-cultural context would not work here. It is important that the community itself gets involved in generating opportunities for health care.

There is a need for some of the following steps that are needed to improve the situation :

Creation of Basic Awareness

Many of the diseases can be prevented or managed by creating awareness about them. Dysentery, cholera, common cold, malaria are some of the diseases which need to be managed and the local health care centres can play a vital role in preventing them from spreading. The schools and the media also have an important role to play. Access to information is a potent tool which can by and large help to curb many of the problems.

Training of Local People

Initiatives such as those undertaken by Barefoot College in generating Barefoot nurses and doctors helps in taking care of many basic ailments. The training of midwives have been responsible for ensuring safety of the mother and the child. The training is essential as in India the ratio of doctors to people is still extremely lopsided. So the community itself can take initiatives in this direction.

Better Traces to Doctors

The condition of government healthcare centres and hospitals in many parts of India is deplorable. The buildings are old and the equipment is not working half the times. The media is full of reports of unhygienic conditions in many of the hospitals. Also the number of patients in OPD in a government hospital are so many that adequate amount of attention seems to be impossible to each patient. So in many situations where the doctor patient ratio is lopsided, the patients are deprived of proper care.

The government can play an important role in changing this. Most of the issues are related with work conditions in the rural areas. Though the government job mandates that the employees have to spend some years in rural jobs, it is a proposition that does not find many takers. Each State government must ensure that there are schools and infrastructure created which would attract manpower from cities to settle there on account of quality of life. The government must make compensation so attractive that people would want to work at such destinations. Rather than giving allowances for working in metros and bigger cities special incentives must be offered to people who are willing to work in rural areas. Allowances for job security or admission of children in professional programmes can change this situation to quite an extent. The government needs to investigate issues at the grassroot level and try and overcome barriers.

Equity Sharing

The community members can come forward and form co-operatives for providing health care services. When stakeholders are themselves involved, then there is a higher probability for things to work out. Also participation of the corporate firms can be envisaged. Some of the experiments as initiated by some hospitals ensure that pharmaceutical firms donate equipments or medicines. There could be some innovative means to overcome such shortages.

Use of Telemedicines Technology

Telemedicine can also be used in areas which can be networked at lower costs. Telemedicine which is the use of connected medical devices in the evaluation, diagnosis and treatment of patients who are situated at other locations. In India, several initiatives have been undertaken in the area of telework and telemedicine. Some of the initiatives are (Radhika, 2001; Srikanth, 2003):

National Telemedicine Technology has been initiated by PGI, Chandigarh; AIIMS, New Delhi; and SGPGIMR, Lucknow. The private hospitals such as Apollo, Escorts Heart Institute and Research Centre have also made commendable strides in this direction. The Apollo group has over thirty-two Telemedicine Centres linking places like Chennai, Hyderabad, Kolkata and Delhi with areas such as Kohima, Guwahati, Aragonda and Sriharikota. The Pune district has also launched an initiative with doctoranywhere.com and Tata Council for Community Initiatives. Such efforts can contribute to lot of cost savings and save lives as well.

REFERENCES

Asia Pacific Centre for Disability, (1991), '*Statistical Data on Disability Profile*' (online) (cited on 28th June 2003). Available from < URL: http://www.apcdproject.org/countryprofile/india/india_current.html >

Disability numbers come down, says NSSO survey, (2004), *Times News Network.*| In-line. Wednesday, April 28.(online). (cited on 28th April 2004). Available from <URL: http://economictimes.indiatimes.com/articleshow/644035.cms>

India's Poor-Findings, '*Analysis, and Options*' Washington, D.C, World Bank

International Institute of Population Sciences (2000), *National Family Health Survey Summary*, India 1998-99. Background Paper 18 Bombay.

Jaipur Foot-Bhagwan Mahaveer Viklang Sahayata Samiti (online) (cited on 5th June 2003). Available from <URL: http://www.jaipurfoot.org/>.

Lakshman, Nikhil (1996), '*Sudha Chandran*' (online) (cited on 5th June 2003). Available from < URL: http://www.overground.be/features/people/sudha.html>.

Lant, H., Wagstaff, Adam (2001), '*India Raising the Sights: Better Health Systems for.*

McCrae, Hamish, (1995), 'Wisdom from the feet up' *The Independent,* 30th March, 1995, UK.

Ministry of Health and Family Welfare. (2000), *Bulletin on Rural Health Statistics in India*. New Delhi: Rural Health Division, Ministry of Health and Family Welfare.

Mukhopadhyay M, ed. (1997), *Report of the Independent Commission on Health in India*. New Delhi: Voluntary Health Association of India Press.

Nalamdana (online) (cited on 15th May 2003), Available from <URL: http://www.nalamdana.org/>.

National Centre for Promotion of Employment for Disabled People (NCPEDP) (online) (cited on 2nd June 2003). Available from. <URL: http://

www.ncpedp.org>

On firmer footing (online) (cited on 7th June 2003), Available from <URL: http://www.chennaionline.com/health/homearticles/jaipurfooting.asp>.

Parikh, Kirti S., Radhakrishnan, R., (eds) (2002), India Development Report 2002, Indira Gandhi Institute of Development Research, New Delhi.

Partners: CHIRAG (online) (cited on 17th May 2003). Available from < URL: http://www.canadaindiavillageaid.com/chirag1987.htm>. Project. Report No. 21751.

Peters, David; Yazbeck, H..; Abdo, S.; Sharma, Rashmi; Ramana R.; Prichett, G.N.V. & Phillips, Tony (2003)

Phillips,Tony (1987), 'New Rural Health Program in Himalayan Foothills', CIVA.

Radhika, V. (2003), *India Abroad News Service*, 2001 http://www.apnic.net

Ray, Arundhati, (2001), '*Giving Visibility to Disability in India*' (online) (cited on 2nd June 2003). Available from < URL: http://www.changemakers.net/journal>.

Rishi Valley-Rural Health Project (online) (cited on 2nd June, 2003). Available from < URL: http://www.ruralhealthrv.org/>.

SEWA Video (online) (cited on 16th May, 2003). Available from < URL: http://www.videosewa.org/aboutus.htm>.

Srikanth, R.P. (2003), Telemedicine: Emergence of the Virtual Doctor, Express Computer Weekly, March 10.

Sulabh International (online) (cited on 14th May, 2003). Available from < URL: http://www.sulabhinternational.org/>.

The World Bank (2001), India Raising the Sights: Better Health Systems for India's Poor Overview, Health, Nutrition, Population Sector Unit, India, South Asia Region.

Tulasidhar, V.B. (1996), Government Health Expenditures in India: Public Financing for Health in India: Recent Trends. A report on work in progress with support from the International Health Policy Program. Mimeo.

TvT Associates.Inc. (1999), India and HIV/AIDS, New Delhi, *The Synergy Project*.

Wang, J., Jamison, D.T., Bos. E., Preker. A., Peadbody. J. (1999), Measuring Country Performance on Health: Selected Indicators for 115 Countries. The World Bank: Health Nutrition and Population Series.

WHO. (2000), *Health a Key to Prosperity: Success Stories in Developing Countries*. Geneva: World Health Organization.

World Bank. (2001), *Project Appraisal Document*. Second National Leprosy Elimination.

6

Corporate Social Responsibility: Selected Interventions

According to Kahane 2004, in today's economic climate, business is increasingly turning towards management practices and investments in low-income communities that benefit both the corporate bottom line and the community in order to achieve competitive advantage, a reality which has been acknowledged by many leading corporate and research agencies. According to the fifth annual Price Waterhouse Coopers Global CEO Survey, 2002, conducted across 33 countries with 1,161 CEO participants, nearly 70 per cent of the CEOs say that corporate social responsibility is "vital" to profitability. This report further says that "while corporate social responsibility has become a near universal value, it manifests quite differently from region to region and from country to country". By referring to the growing gap between poor and the rich countries, the digital divide, the environmental responsibility of the industry, the social impacts of the corporate strategies and investments, and the need for crafting of more stable global economic systems, this report reemphasizes that while corporate social responsibility is perceived as a strong connection between corporate profitability and societal improvements, including infrastructure, education and addressing the ravages of AIDS in Africa and other developing nations, it is more of a short hand for environmental responsibility ethical economics or charitable giving in Europe, America, etc. Interestingly no one denies that Corporate Social Responsibility (CSR) is all about doing the right thing and creating a positive social reputation of the company. Companies and their CEOs the world over equate corporate special responsibility with social reputation of the company as can be seen in the Figure 6.1 that though their definitions of CSR differs their concept of the company reputation remains the same.

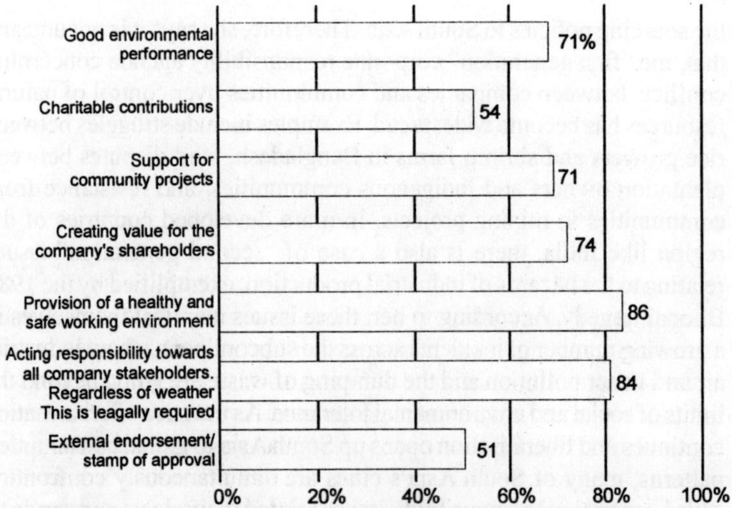

Figure 6.1 : Corporate Social Responsibility and Factors influencing Company's Social Reputation

Source: Price Water House Coopers CEO Survey, 2002.

Desai, 2002 (PWC. online) while acknowledging these differences says, "In my experience, Europeans have greater sensitivity to emissions issues, while the focus in Latin America is on resource impact issues. I would say that social issues—the impact on the population of industrial pollution, questions about welfare of the minority groups, etc., are probably stronger in Asia. However, there is a commonality and one of the most powerful causes is that corporate managers are part of a community of corporate managers around the world and today they admire the manager who successfully combines shareholder values with environmental and social responsibility". More and more managers are now willing to concede that an idea which is beneficial to the company need not be well received by the environment in which the corporate hopes to operate. This is especially true for organizations hoping to expand into the largely unexplored South Asian region. This is an area which is characterized by large population, qualified labour, abundant natural resources and untapped markets. Therefore, as they gradually narrow down to the South Asian region. Kumar (2004) has pointed towards the fact that numerous North American and European companies are increasingly shifting their global operations to their subsidiaries in emerging economies along with their standards towards the sourcing policies to South Asia. Therefore, she says it is not unusual that, the "first generation" corporate responsibility agenda concerning conflicts between companies and communities over control of natural resources has become widespread. Examples include struggles between rice growers and shrimp farms in Bangladesh, land disputes between plantation owners and indigenous communities, and resistance from communities to mining projects. In more developed countries of the region like India, there is also a case of "second generation" issues relating to the hazards of industrial production, exemplified by the 1984 Bhopal tragedy. According to her, these issues manifest themselves in a growing number of incidents across the subcontinent, where industrial air and water pollution and the dumping of waste are going beyond the limits of social and environmental tolerance. As the pace of urbanisation continues and liberalisation opens up South Asia to global consumption patterns, many of South Asia's cities are simultaneously confronting 'third generation' responsibility issues related to products and services, whether autopollution norms or pesticide residues in mineral water. What gives an added twist to this threefold challenge of corporate responsibility is the extent to which local and global dimensions are deeply interpenetrated.

Just as the definitions of CSR vary from corporation to corporation, so do the interventions. In a recent Corporate Social Responsibility Report by *Business Ethics*, 2004, the top 100 companies of the USA expressed their CSR in many different ways. Given below are two examples from that report—

Bringing in top honours in service to overseas stakeholders were two firms: No. 1—ranked Fannie Mae. It has created a Native American Conventional Lending Initiative to help finance $75 million in loans on trust land for the Navajo Nation of Arizona, Oneida Nation of New York, and Menominee of Wisconsin.

Coming in No. 2 on the list this year was Procter & Gamble, which excelled in service to minorities and women, and to the community. The firm has donated to help disadvantaged youth in Vietnam, to combat childhood malnutrition in India, and to provide earthquake relief in Turkey. "Over 2 million children in developing countries die each year from water-borne diarrhoeal diseases," said P&G spokesperson Terry Loftus, "We have developed a technology that allows people in the developing world to clean and disinfect water in their homes at low cost".

"We have a slogan around here," says Robert W. Lane, chairman and CEO Deere & Company. " 'No smoke, no mirrors, no tricks: just right down the middle of the field.' That's John Deere." (John Deere, Midwestern maker of farm machinery. Ranked No. 6 among the 100 Best in 2004).

Fuel cells cannot only "supplement our ageing electricity grid, they have the potential to replace the polluting internal combustion engine". (Mark Baffa, Director of Fuel Cell Products, Modine Manufacturing of Racine, Wisc. (Ranked No. 32),

Rochester, N.Y.-based Eastman Kodak does with its leading-edge anti-discrimination policies toward gay, bisexual, and transgender employees. These policies have earned the company the rare distinction of receiving a perfect score from the Human Rights Campaign, an organization working to end sexual orientation discrimination.(Ranked No. 58 overall—tied for highest score in service to employee).

Way back in 1909, Gandhiji in *Hind Swaraj* stated, that "the civilization of the West then being welcome in India in the name of modernity is '*Bhasmasur*' (destructive monster). It is a civilization that equates consumerist lifestyle and abundance with development (Sheth, online). It is astonishing how his perceptive insight—"the Earth has enough to satisfy the need of all the people, but not for satisfying greed of some" has become the rallying point on the Agenda 21 of the Earth Summit in Rio de Janeiro. Gandhi's 'trusteeship' model offered distinctive approach to the questions of ownership and duty. In other words, he was of the opinion that mankind of which he was a small part was but a custodian of the natural wealth and therefore had no moral right to destroy it or use it selfishly. Hence his emphasis on adoption of more labour intensive techniques, a concept that formed an intrinsic part of the 1990 World Development Report, and minimum utilization of non-renewable resources, that to say that he advocated an "Economy of permanence".

Corporate responsibility is thus not something new or somehow foreign for South Asian business culture. And perhaps this acted as the guiding light for many corporate houses of the emerging India of nineteenth century. That is to say that the CSR policies of various firms was of a different nature in the Indian sub continent. Kumar (2004) has quoted the famous Jamshedji Tata, the founder of Tata Steel, from Lala (1981), as "We started on sound and straightforward business principles", declared the chief executive at the opening of his new textile factory, "considering the interests of the shareholders, our own, and the health and welfare of the employees the sure foundation of our prosperity....". Speaking in 1895, Jamshedji Tata was not only one of the pioneers of industrial development in South Asia, but its first proponent of corporate responsibility. From the beginning, workers at Tata Iron & Steel Complex at Jamshedpur, opened in 1911, worked an eight-hour day, when factories in Europe and North America still had a minimum of 10- or 12-hour a days. Tata also commissioned social audits of the company's community impacts long before they became fashionable in the West. There have been various pro-active measures taken by Indian corporate houses as regards CSR. Given below are few example taken from the CSR Survey 2002 to exemplify some of the Indian initiatives in this regard:

According to Y.S. Navate, Senior Manager (Projects)

Gujarat State Petronet Limited "As we operate in a hazardous industry—Health, Safety and Environment Management is crucial for our business, in addition, our facilities are spread all over Gujarat and no amount of control could ensure safe management of these utilities without our linkages with and awareness of community to guarantee safe operations".

Setting New Standards at Orchid Chemicals' vision articulates creation of sustainable value as the principal goal and has a set of core values for the attainment of the said goal. The core values include:

1. Respect for human being and a premium on intellect,
2. Quest for international standards in all facets,
3. Customer satisfaction through quality and service,
4. Harmony with environment, and
5. Maximization of stakeholder value.

In keeping with the core values Orchid has established systems, dedicated resources to take various initiatives.

Shri G.D. Birla, the founder, of the Aditya Birla Group brought succour to the suffering through extensive charitable work. The late Mr. Aditya Birla initiated a paradigm shift when he stated that charity alone cannot be a long term solution. Instead, he said, "We should help people in a way that they are able to stand on their own feet and earn money continuously. In this way, their livelihood is never at stake". His philosophy is best articulated in the following statement. If you give a hungry man fish one day, he will eat it and the next day he will be starving again. Instead, "if you teach him to fish, he will never go hungry during his lifetime". The group believed in the trusteeship concept of management, which entails ploughing back part of its profit into community. Mr. Kumaramangalam Birla, has enlarged the group's CSR commitment by adopting a Triple Bottom Line approach to management where accountability lies at the heart of the concept as it involves factoring holistically the interests of all stakeholders—shareholders, customers, employees and the community at large.

As one of the largest financial institutions in India, ICICI's overall mission has been to build the capacities of commercial entities and, thereby, enable them to become agents of economic change. We believe that building the capacities of the poorest of the poor to participate in the economy and society is a natural extension of this role. With this in mind, ICICI has been contributing to the social sector for nearly 30 years, mainly by financing and advising organisations on their working.

The Indian perception of corporate social responsibility, has been summarized by the Corporate Social Responsibility Survey 2002-India, in the chart below:

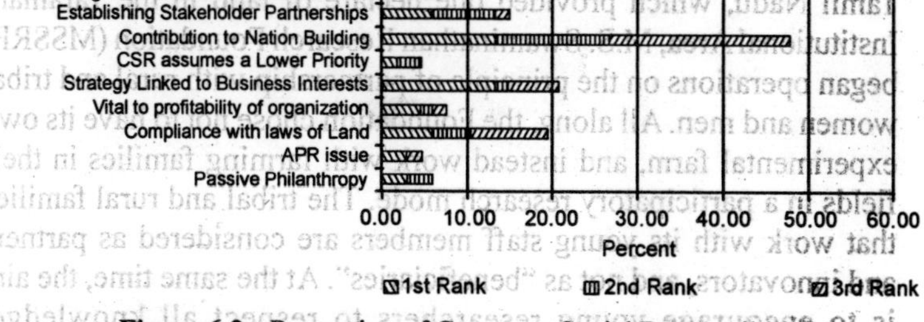

Figure 6.2 : Perception of Corporate Social Responsibility

Source: CSR Survey, 2002-India.

The diversity in definition of CSR is amply reflected in the responses received from companies on their perception of CSR and is very similar to various responses received in the Global CEO report, and that there is a near unanimity that CSR is very much included in the domain of corporate action and that passive philanthropy is no longer sufficient in the realm of CSR.

M.S. Swaminathan Foundation

This is an organization that based its operating ethos on the bedrock of Gandhian philosophy of sustainable rural development. Perhaps because its founder Prof. M.S. Swaminathan has deeply felt the need to integrate the sustainability of the rural development with the ecosystems that surround it. Unlike many others, he could not be a mute spectator to the increasing degradation of the basic life support systems of land, water, forests, biodiversity and atmosphere. He grew concerned by the increasing poverty as well as social and gender inequity where rapid growth in human population resulted in reduced per capita availability of land and water. Nor was he able to come to terms with the directionless technological development, which coupled with high rates of unemployment, resulted in lopsided economic growth. So instead of waiting for things to happen, he set up the M.S. Swaminathan Foundation, an autonomous non-profit trust, that was registered in 1988 at New Delhi. Initial funds came from the World Food Prize which Dr. M.S. Swaminathan received in 1987. It aimed to organize research and training designed to promote a job-led economic growth strategy in rural areas. However, the Foundation moved base to Tamil Nadu in 1989 because it was considered an ideal location to start some of its initial operations like those relating to coastal ecosystems. Here with the assistance of the government of Tamil Nadu, which provided one hectare of land in the Taramani Institutional Area, M.S. Swaminathan Research Foundation (MSSRF) began operations on the principle of partnership with rural and tribal women and men. All along, the Foundation chose not to have its own experimental farm, and instead work with farming families in their fields in a participatory research mode. The tribal and rural families that work with its young staff members are considered as partners and innovators, and not as "beneficiaries". At the same time, the aim is to encourage young researchers to respect all knowledge, irrespective of its origin. It is this genuine partnership based on mutual esteem, between those whose wisdom grows from real life experience and toil in sun and rain and those with advanced academic training, that has given the Foundation its scientific strength. All these factors led to the Foundation defining its research agenda in terms of **sustainable development**, which is rooted in the principles of ecology, social and gender equity, employment generation, and economic viability. The fostering of a pro-nature, pro-poor, pro-women, and pro-employment orientation to technology development and dissemination in rural areas has thus become MSSRF's mission. MSSRF's approach has been to enlist appropriate blends of traditional and frontier technologies as allies in the movement for economic and ecological well-being and gender equity, so that the technology which was instrumental in the past in increasing economic and social disparities and causing ecological harm, could be harnessed for a reversing the harms caused to the ecosystem.

Soon MSSRF, sought to lay down the foundation of integrated natural resources management in order to increase the crop productivity in the field of agriculture. It became evident that MSSRF's goal was one of spreading an evergreen revolution based on sustainable advances in biological productivity, well in line with its priorities:

1. The conservation and enhancement of natural resources—particularly land, water, and biodiversity,
2. The promotion of sustainable and equitable agricultural and rural development,
3. The development of ecologically sound and economically and socially viable 'eco-technologies' by integrating traditional and frontier technologies,
4. The generation of greater opportunities for skilled employment, particularly for rural women and youth, and
5. Reaching the unreached and voicing the voiceless in terms of technical-literacy and gender equity.

Designed to converge on this goal of promoting a pattern of agricultural and rural development, MSSRF has focused its activities in the five major programme areas :

1. Coastal Systems Research,
2. Biodiversity and Biotechnology,

3. Ecotechnology and Food Security,
4. Gender and Development, and
5. Informatics.

In terms of sustainable human welfare, MSSRF hopes that these scientific goals would lead to a synchronized amplification of the environmental safety of rural areas and the livelihood protection of rural families. It has always been a strongly held belief at the Foundation that the notion of sustainability has to be a dynamic one, leading to a continuous improvement of biological yield on an ecologically, economically, and socially sustainable basis.

Another aspect that marks out MSSRF from other such organizations is, that right from the beginning, MSSRF was consciously designed as a scientific and training institution "without walls", i.e., an institution which relies on partnerships for its vitality and effectiveness. Not surprisingly MSSRF has fostered strong inter-linkages with other scientific and corporate institutions, both national and international, for developing and testing front line sustainable rural technologies. In the academic world, the Foundation is recognised as a Post-Graduate Research Centre by the University of Madras, Anna University, Chennai, and the Osmania University, Hyderabad and the resultant output is over 10 doctoral thesis and numerous research papers and technical publications, besides monographs and reports of Dialogues and Conferences. Some of the other interesting outcomes and highlights of all this academic/research and developmental work are :

1. The Deustche Gesellschaft Fur Technische Zusammenargeit (GTZ) of Germany funded a project on Traditional Conservation of Biodiversity by Tribal Communities.
2. The International Atomic Energy Agency (IAEA), Vienna, assisted a project on Design of Database on Intra-specific Variations in Plant Species of Mangrove Ecosystems.
3. The Netherlands Government financed the organising of an Agro-Biodiversity Conservation Corps.
4. The Global Environmental Facility (GEF) of the United Nations requested MSSRV to prepare a report on Agro-Biodiversity Conservation and Sustainable Use.
5. SDC sanctioned an assignment on Establishing and

Disseminating Multimedia Database on Agro-biodiversity and its Utilisation by Farm and Tribal Communities as a component of the Resource Centre for Farmers' Rights.

6. The Department of Scientific and Industrial Research (DSIR) of the Government of India asked the Foundation to undertake a study for National Mapping of Science using databases under the Informatics programme of NISSAT.
7. On the recommendation of the Government of Tamil Nadu, a long-term project was approved by the Council for Advancement of People's Action Technology (CAPART) for establishing a Technical Resource Centre for Rural Technology at MSSRR.
8. The Bernard Van Leer Foundation supported a project for advocacy, networking, training, communication, and resource materials development on issues related to the young child, including strengthening of a non-governmental network to advocate the cause of the young child.

The Information Technology revolution has created numerous opportunities of providing low cost information to 'unreached' masses of the country. Currently as a developing nation, India suffers from unequal distribution of scientific and developmental information. This aspect is further compounded by absence of databases of local knowledge and practices, which can combine with ICT's ability to transform generic information to suit local needs. Many prominent Indian social scientists are of the opinion that there is an urgent need to document indigenous knowledge and traditional agricultural practices thereby avoiding the risk of endangering them. Rural families and marginal farmers can use this information and hope to improve productivity of labour and inputs. Even under such circumstances, ICTs have been at the vanguard of spreading literacy and application of rural development programmes among the vast underprivileged population of the country.

One such project "The Information Village Research Project" was launched in 1998 in Pondicherry, by M.S. Swaminathan Research Foundation as an experiment in electronic knowledge delivery to the poor, to bring information age in the form of communal telephone and Internet access to the rural populace of the region (www.mssrf.org). This project had a clear cut goal that "India's network must reach the villages if it is to make a meaningful contribution to the quality of life", so that a model of information diffusion and exchange in rural areas using complex information and communication technologies could be developed.

Conclusion

The discussion in this chapter points out at the importance of transparency for the firms and wide range of interventions which can be instituted the corporate firms and different NGOs have district roles to play. The firms need to add value to various stakeholders. Society is an important stakeholder in a firm in the form of both the demand for the products or services and also as employees, suppliers and the financial investors. There is a need for concern towards environment. The corporate firms are also entrepreneurial initiatives which generate employment and add value to the basic resources. If firm is not transparent in its operations, there are bound to be issues which would lead to a scenario wherein the firm does not do well and wastes the resources of the society and leads to unemployment as well. Transparency and the disclosure of information forces the firms to perform well and ensure productivity as well. Internationally there is a conscious movement towards the scenario wherein there is a pressure on board of directors and disclosures of financial information. On the other hand, organizations, Swaminathan Foundation operate in a wide range of areas ensuring that they contribute to a holistic development and the same time have their operations profitable.

REFERENCES

Business Ethics (2004), 'Corporate Social Responsibility Report, 2004', *Business Ethics,* 18:1, Spring (online)(cited on 21 July 2004). Available from <URL:/ / http://www.business-ethics.com/current.htm>

Kahane, Michele (2004), "*Win-Win Business Strategies for Companies and Communities*" Ford Foundation (online). (cited on 14th July, 2004). Availablc from URL:http://www.knowledgecentre.com

Kumar, Ritu (2004), Corporate Responsibility: The Emerging South Asian Agenda, Seminar series, British Council, (online) (cited on 14th July, 2004). Available from <URL: http://www.britishcouncil.org/csr2004-seminarseries.pdf>

M.S. Swaminathan Research Foundation (online) (cited on 10th July 2004).

Available from <URL:http://www.mssrf.org>

Price Waterhouse Coopers, CII, British Council & United Nations Development Programme, "Corporate Social Responsibility Survey 2002-India', (online) (cited on 23 July 2004). Available from <URL: http://www.indiapartner shipforum.org/csrs2002/CSR per cent20Report per cent202002.pdf>

Price Waterhouse Coopers, "Uncertain Times, Abundant Opportunities", 5th Annual Global CEO Survey, 2002 (online)(cited on 23 July, 2004). Available from <URL:http:// pwcglobal.com>

Sheth, Pravin (2003), 'The Eco-Gandhi and Ecological Movements' (online) (cited on 26 May, 2003). Available from <URL: http://www.mkgandhi.org/ articles/ecogandhi.htm>

7

Role of Corporate Sector in Sustainable Development: Lessons from Success Stories

BACKGROUND

The pictures presented by poverty are grim. By 1999, 1,170 million people lived in extreme poverty[1] (World Bank, 2003).

"Though the poverty levels have declined in East Asia and Pacific, In Sub-Saharan Africa, the number of people steeped in extreme poverty increased by 74 million. For 1998-2000, the FAO estimates that 799 million people or 17 per cent of the population in developing countries were undernourished. Additionally, there are another 30 million undernourished people in transition economies of Eastern Europe and the former Soviet Union.115 million children remain out of school. In South Asia, where only 61 per cent of girls complete primary school, the average woman has 3.4 years of schooling. In 2002, 42 million adults and 5 million children were living with AIDS. Lack of clean water and basic sanitation is the main reason for so many health related problems. In 1990, diarrhoea led to 3 million deaths, 85 per cent of them were children. In 2000, 1.2 billion people still lacked access to an improved water source."

—Human Development Report, 2003

In a world, where there are amazing technological achievements at one end, there are issues of people living in abject poverty. The idea is to include people at the bottom of the pyramid and include them in the economic development process. The corporate sector has a distinct role to play in bringing out this change. The poor pose a lot of challenge for the public sector. The public sector has neither the expertise nor the resources to provide goods and services on a scale sufficient to reach the approximately four billion people who currently earn less than $ 2,000 per year. The private sector has both. Prahalad and Hammond (2004) in their study point out that there are some innovative initiatives undertaken by the private sector which aims at products and services targeted at the poor (bottom of the pyramid). In their study, they have pointed out the following:

1. The poor live in very high cost economies. For most products and services, they end up spending more money for most products that they buy. Costs to the poor can dramatically be reduced if they could benefit from the scope, scale and supply chain organization of large enterprises, as do their middle class counterparts.
2. Many interventions end up exploiting poor rather than do them good in the long run.
3. The poor have a purchasing power.
4. Many of the BOP markets are geographically concentrated. A slum such as the one located in Dharavi in just 175 hectares, generates $450 million in manufacturing revenues.
5. The poor welcome new technologies.

Sustainable Development and Profitability

The World Commission on Environment and Development (1987, p. 8) states, "sustainable development meets the needs of the present without compromising the ability of future generations to meet their own needs". Reinhardt (2003) indicates that there are issues in measuring sustainability at the firm level and there is also confusion about the term sustainable development itself. When one talks about the term sustainable development at the firm level then there are issues about its measurement. Reinhardt (2003) quotes the two DuPont's formulas about measuring the sustainability. One is the private cost test and the other is the social cost test. Under the first test, a sustainable firm is one that augments its shareholders value when all stocks and flows are measured at the prevailing private costs, as those costs are determined by market forces and by current government regulations. Companies deliver profits to their shareholders by creating value (Holliday, 2001). Under the second test, a sustainable firm is one that creates value when all of its costs and revenues are measured at their social costs. This test is exactly analogous to the national level sustainability requirement. To be sustainable from a social cost perspective, a firm must be creating value and reinvesting enough of that value to maintain its capital stock at an undiminished level. As was true for national economies, to test for a firm's sustainability, it is essential to consider conventional accounting measures as well as environmental performance. If a company has to pass this test, it also delivers profit that are measured, not according to social costs but by market prices.

For a corporate firm, profitability is important for survival and existence. The challenge is about being able to identify the right kind of business product which not only creates a lasting value for the organization but also involves people at the grassroots who have not been included in the development process[2]. The world statistics indicate that poverty has been a curse for mankind which deprives people of self esteem, empowerment and denies basic rights. The corporate world has a huge capacity to initiate meaningful initiatives for the human beings. There have been a few experiments in India which demonstrates a faith in the power of networking and the capacity of shaping their own destiny by some communities. There are diverse initiatives by NGOs, the corporate world and the support by the government.

The objective of this chapter is to analyse the different interventions initiated by the corporate firms and NGOs which help the development of communities living in poverty. The objective is to identify interventions which can be replicated, are sustainable and can help the underdeveloped regions to move to a level of economic prosperity. The study also assesses the role of the corporate sector and MNCs. The objective is to develop frameworks which could work in areas with different levels of development.

Each country has a unique historical background and a culture which needs to be preserved. Blind imitation of western models for initiating development is not healthy as that replicates the same culture diluting the individualistic characteristics. This research uses the multiple case study approach which reflect the participative approach directed to alleviate poverty and empower people at the grassroots level. In all these cases, the corporate sector has had a distinct role to play. These are interventions which have been initiated and are targeted at various stages of development. The nature and extent of intervention varies as it moves from lower stage of development to a higher stage of development. The development in a country could be sustainable if strategies are aligned to the resources available in abundance. This chapter proposes models for development and links the nature of intervention with the stage of development of that geographical area. The chapter also makes policy recommendations for different category of interventions.

Corporate Interventions

Competitiveness is important for a nation's prosperity (Porter, 2003). The framework developed by Porter (2003) links competitiveness: quality of factor conditions, context of firm strategy and rivalry, the quality of local demand conditions, and the presence of related and supporting industry. However, the low income countries could shorten their learning curve and could leapfrog to the status of developed countries. This could be done by establishing linkages with local strengths. India has witnessed some of these unique experiments which bring to light some interesting display of innovation and strengths at the grassroot level. These are communities which found tremendous strength by identifying their own strengths and utilising them to create brands which have given the multinationals a run for their money.

The ITC e-Choupal Initiative

It is an experiment which demonstrates the identification of strengths that exist at village level, identification of a methodology which results in higher incomes, use of information technology which creates empowerment and redefines the roles of individuals. It also demonstrates a unique innovation on the part of the corporate firms to redefine the supply chain management practices which contributes to added value.

ITC is an Indian diversified firm which operates in the area of tobacco, hotels, consumer products and agri-based products. The corporate aim was :

1. To become India's largest agri-produce processor servicing 1,00,000 villages covering 10 million farmers by 2007,
2. Create information superhighway to rural economy, and

3. Single point of contact between farmers and a range of suppliers of agri inputs and consumer goods—Monsanto, Eicher, Nagarjuna Fertilizers.

The Indian Backdrop

One of the interventions suggested to alleviate poverty is to increase the productivity of small farmers. 70 per cent of the world's poorest farmers live in rural areas and depend on agriculture (Human Development Report, 2003). A typical village in India does not have basic amenities such as electricity and water. Illiteracy levels are high. However, the same group became adept at using the e-commerce which changed their lives. It is interesting to see also at the same time how the rural market in India behaves with regards to the demand for consumer goods. Table 7.1 gives an height on how much population resides in how many villages.

Table 7.1: Distribution of Villages in India

Population	*No. of Villages*	*Per Cent of Total Villages*
Less than 200	114,267	17.9
200-499	155,123	24.3
500-999	159,400	25.0
1,000-1,999	125,758	19.7
2000-4,999	69,135	10.8
5,000-9,999	11,618	1.8
10,000 and above	3,064	0.5
Total number of Villages	638,365	—

Source: Kashyap, P. (2003) *The Marketing Whitebook, 2003-04*, Business World, Kolkata.

In agricultural products, there are *mandis*, which are agricultural markets set up by the State Governments to procure agricultural produce directly from the farmers. Located in high production centres of different crops, these markets can be categorised as grain *mandis*, cotton *mandis*, soya *mandis* and so on. There are about 6800 *mandis* in India. The average population catered to by each *mandi* is 1.36 lakhs. Most agricultural areas with population more than 10,000 have *mandis*.

Annamalai & Rao (2003) have criticized the *mandi* system stating that the *mandi* system does not serve the farmer well and is burdened by inefficiency. Depending on the geographic proximity, the farmer takes his goods to the nearest *mandi*. The system is viewed as based on exploitation. There are multiple points before the supplies are bought that contribute to wastage. We see that in Table 7.2 that the number of villages and *haats* to cater to them are limited. In HP, 10 *haats* are catering to an average of 1983 villages. This means that it is extremely difficult for a farmer to travel large distances to sell their produce at a *haat*.

Table 7.2: Villages and *Haats* in India

State/UT	*No. of Villages*	*(No. of Haats)*
Jammu & Kashmir	6652	(NA)
Himachal Pradesh	19,831	(10)
Punjab	12,729	(16)
Chandigarh	24	(0)
Uttaranchal	16,805	(NA)
Haryana	6,955	(0)
Delhi (rural)	165	(28)
Rajasthan	41,353	(261)
Uttar Pradesh	107,440	(14,121)
Bihar	45,113	(10,681)
Sikkim	452	(19)
Arunachal Pradesh	4,065	(NA)
Nagaland	1,315	(46)
Manipur	2,391	(7)
Mizoram	817	(49)
Tripura	870	(736)
Meghalaya	6,023	(319)
Assam	26,247	(4,044)
West Bengal	40,783	(456)
Jharkhand	32,615	(NA)
Orissa	51,352	(3,887)
Chhattisgarh	20,308	(NA)
Madhya Pradesh	5,392(	NA)
Gujarat	18,544	(NA)
Daman & Diu	23	(0)

(Contd.)

Table 7.2 (*Contd.*)

Dadar & Nagar Haveli	0	(6)
Maharashtra	43,722	(3,578)
Karnataka	29,483	(1241)
Andhra Pradesh	8,183	(NA)
Goa	359	(8)
Lakshadweep	24	(0)
Kerala	1,364	(670)
Tamil Nadu	16,317	(1169)
Pondicherry	92	(0)
Andaman & Nicobar	547	(0)
India	6,38,365	

Source: Marketing Whitebook *op. cit.*

Table 7.3 gives insights into Regulated Agri Produce Markets in India.

Table 7.3: Regulated Agri Produce Markets (*Mandis*) in India

State	*No. of Mandis*
Andhra Pradesh	23
Assam	35
Bihar	828
Goa	5
Gujarat	30
Haryana	75
Himachal Pradesh	29
Karnataka	44
Kerala	5
Madhya Pradesh	84
Maharashtra	827
Orissa	123
Punjab	667
Rajasthan	384
Tamil Nadu	270
Tripura	21
Uttar Pradesh	645
West Bengal	456
Chandigarh	3
Delhi	17
Pondicherry	5

The Business Model

The model is centred on a network of e-chaupals, information centres equipped with a computer connected to the internet located in rural villages. ITC appointed *sanchalaks* (co-ordinator) and *samyojaks* (collaborator) into the system as the provider of logistics support. Each e-choupal would serve about 10 villages within a five km radius. There are a million farmers being served in nearly 11,000 villages through 2000 e-choupals in four States (Madhya Pradesh, Karnataka, Andhra Pradesh and Uttar Pradesh). The traditional model in indicated in Fig. 7.1.

Traditionally, the soya or wheat crop would be traded at *mandis*. Farmers were exploited by traders, who often bought the wheat at low prices claiming it to be of poor variety. Storage, transport and handling facilities were limited. So intermediaries like traders were needed. Intermediary margins were high, accurate market signals were non, existent, wastage rampant and processing yields low. Conventionally, a typical farmer would sell to small trader called the *kaccha adat*. This man in turn sold it to a larger trader called the *pakka adat*. From here the produce found its way to the local *mandi* where a large trader came into picture. Going through a loop like this meant procurement costs were as high as Rs. 700 per tonne of soya. In addition, there were losses in transit and tax burden as well.

Brokers compare these statistics with the American mid-west market. Its farmers produce 36 bushels of wheat per acre, one of the highest yields per acre and its millers achieve flour extraction levels of 75 per cent. Average waste levels are 2 per cent which are lowest in the world and well ahead of the 8-11 per cent in India. Wheat farmers in the US now get 92 per cent of the delivered mill price as opposed to less than 70 per cent that Indian farmers now take home. ITC started the process of disintermediation. So, by providing the information access to the farmer rather than the intermediary, the value is added directly.

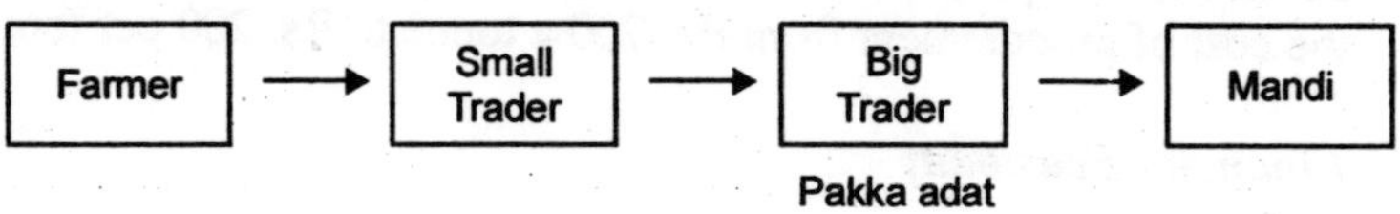

Procurement Cost: Rs. 700/tonne

Fig. 7.1 : Conventional Channel for Selling Soya

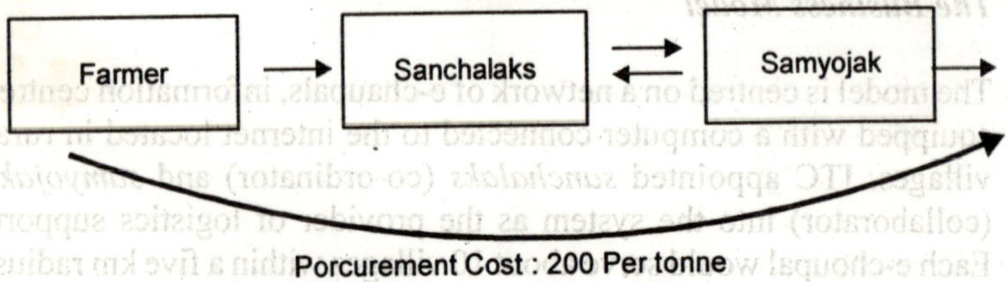

Fig. 7.2 : The ITC Model for Soya Procurement

So ITC identified co-ordinators at the village level who were called *sanchalaks* who used the net to provide spot quotes after examining produce sample. If the price was right, the farmers would take the produce to the ITC collection centre. The *samyojaks* were the collaborators who would co-ordinate the group of villagers. He would perform the documentation work. He would supply farm inputs from ITC and partner companies to *sanchalaks*. They also build relationships with *sanchalaks* and farmers. The ITC model is indicated in Figure 7.2.

When ITC started the experiment it was the middle of 1990s. ITC first leased few centres and started scouting for villages around these centres. The computer was placed at the *sanchalak's* house who would head each choupal. He was trained on using the computer. He was carefully chosen so that farmers could trust him – neither too big which would distance him and neither too small so that he may not be respected. ITC started to pump information on *mandi* prices through the internet into the *sanchalaks* home. Information on best farming practices and weather forecasts were also made. The *sanchalaks* were paid 0.5 per cent for each tonne of soya that originated from their choupal. There were commission agents in *mandis* whose role was redefined by ITC. They were called *samyojaks* or co-ordinators. They would use their ties to nominate *sanchalaks* but also do the relevant *mandi* documentation. For villages situated far off, the *samyojaks* would aggregate the grain and bring it to ITC. For this he was paid commission of 1 per cent. ITC as a result reduced the cost of procurement from Rs. 700 a tonne to Rs. 200 per tonne.

Financial Feasibility

On an average it cost Rs. 40,000 to set up a basic *choupal*. In places where connectivity was poor ITC would have to invest in VSATs. These investments escalated the costs to Rs. 1 lakh. In MP alone, when the experiment started ITC had 1,045 e-choupals spread over 6,000 villages serving over six lakh farmers. This was for soya crop which was a 5 million tonne crop in MP. Wheat was 14 times bigger.

ITC Gameplan

There would be 1.5 million outlets in the country. There would be an army of mobile traders and cycle distributors. There would be a huge earning on transaction fees for every deal that is taking place. The market for agricultural products is valued at Rs. 1,75,000 crores. The tobacco market is just Rs. 15,000 crores. So it makes a lot of business sense to diversify the risks and add new business areas which also contribute not only to the bottom lines but also to the profitability as well. The investment made by ITC would be recovered in five years for wheat procurement. The infrastructure created could be used for other purposes as well. It also sets up a relationship at the grassroot level. This could in years to come open doors for more revenues to come in as it would also expose the community to new products as well.

The ITC had tied up with Monsanto for seed selling. The FMCG's and consumer durables could be delivered through the same network.

The investments in the e-commerce initiatives are different as compared to the normal manufacturing operations. The cost curves for e-based businesses behave differently as compared to the manufacturing set-up. Choi *et al.* (1997) and Choi and Whinston (2000) discuss about investments in the e-businesses. The total cost curves of many physical products and services are U-shaped. First as the quantity increases the cost declines, but later the cost increases due to the growth of both the fixed and the variable costs (especially administrative and marketing costs). In contrast with digital products, the variable costs per unit is low and almost fixed, regardless of the quantity. Therefore, cost per unit will decline as quantity increases due to pro-ration of the fixed component of the cost over more units. This results in increasing returns with increasing sales (Turban, 2002).

Most FMCGs currently operate where they are viable while servicing populations above 1,00,000. There are enormous pockets of widely dispersed markets where population sizes are below 2000. ITC with 1.5 million outlets across the country, an army of mobile traders and cycle based distributors claim to understand the nuances of catering to these populations. By collaborating with ITC, a potential seller does not have to invest in infrastructure. ITC simply takes a small transaction fee for every transaction for every deal taking place on its network. The market for agricultural markets is Rs. 1,75,000 crore out of which ITC only accounts for Rs. 15,000 crore.

Comparison with Traditional Models

The agribusiness initiatives taken by some of the other farmers have not been successful. Initiatives by some of the well known companies such as Mahindra & Mahindra, Tata Chemicals, EID Parry, Rallis and Nagarjuna Fertilizers took up agribusiness ventures by setting their own models. Most of them faced problems for growth. The "*Shubh Labh*" venture started by Mahindras had set up 36 centres in 10 States by the end of 2002. Even ikisan.com (Nagarjuna Fertilizers) and Indiagriline.com and Parry's corner (EID Parry) which tried to create an e-market place could not scale up.

Some of the problems with these models were:

Firstly, there was a lack of trust in the farmer as these models depended on the intermediaries who in turn were not trusted by the farmer.

Secondly, these models had high fixed costs. The fixed cost of a Rallis Kendra was Rs. 0.5 million to 0.6 million, whereas the variable cost came up to Rs. 1 mn (this included the cost of a PC, telephone, furniture, 2-3 motorcycles, for field staff, rentals and salaries). Rallis covered 20,000 acres but could not scale up. Every centre had 6-8 employees. For 10 centres, 60 people were to be employed. Taking an average salary of Rs. 0.15 million would mean an outflow of Rs. 100 million/month in salaries itself. Unless volumes go up, operating costs would be high.

Thirdly, there were issues of channel conflict. Also many companies did not have enough knowledge in procurement. Corporations need the organizational knowledge of handling samples. The market does not reward you for scaling but for taking positions.

From the organization's perspective, profits do not come easy in agribusiness unless the organization has sufficient control over the supply chain of the agricultural produce. In Madhya Pradesh the farmers were located in remote villages scattered throughout the State. Next issue before the organization was that of the middlemen for obtaining good quality products. The middlemen exerted influence on the farmers by concealing the prevailing market prices and other related information. They made large profits for themselves by blocking access to such information.

On the farmers side, there were issues such as they were trapped in a vicious circle of low investment, low productivity, weak market orientation, low value addition, low margin, low risk taking ability. So the paradoxical condition that despite abundant natural resources, Indian farmers remained globally uncompetitive. This was evident in Indian performance on the global scale. The immediate aim for the e-chaupal initiative was to integrate ITC's association with rural suppliers as well as to develop new markets for its own and third party goods.

The rural symphony started in Madhya Pradesh which was dominated with firms such as Ruchi Industries. ITC soya chaupals are all over Madhya Pradesh and cater to more than 6000 villages. The vision is to get farmers better price for their crops.

TARAhaat – A Case Study

This case study highlights an attempt to achieve connectivity for the poor. *TARAhaat* comprises a commercially viable model for bringing relevant information, products and services *via* the internet to the unserved rural market of India, from which an estimated 50 per cent of the national income is derived. *TARAhaat* is promoted by Development alternatives and its alliance partners include Hughes Escorts Communication, KLG Systel, jaldi.com, James Martin, Hewlett Packard, Oracle and the Global Development Gateway sponsored by the World Bank and the Gates Foundation. Excelsior Ventures Management and James Martin & Co. are providing initial equity capital together with management and operational resources. *TARAhaat* is a business enterprise of Development Alternatives (DA), an NGO focused on sustainable rural development in India, and its marketing arm, Technology and Action for Rural Advancement (TARA). Tarahaat uses a franchise-based business model to bring computer and internet technology to rural regions and plans to use these technologies to create revenue streams leading to financial viability for itself and its franchisees. These business objectives are balanced with the social objective of uplifting rural India through easy access to relevant local and global information and propelling it into the twenty-first century. *TARAhaat* also hopes to find the right balance of rural and peri-urban franchises to be able to subsidize rural locations with lower potential.

The initiative includes the development of a mother portal, "TARAhaat.com" supported by franchised networks of village cybercafes and delivery systems to provide a full range of services to its clients. The portal has subsidiaries:

TARAdhaba	:	provides connectivity and access to new world.
TARAbazzar	:	provide access to products and services needed by rural households, farmers and industries.
TARAvan	:	deliver goods.
TARAdak	:	connect the rural families to the daughter married far off and the son posted on the front.
TARAguru	:	decentralized university providing mentoring and consultancy to village based mini enterprises.
TARAcard	:	enable the villager to order goods and services on credit.

The effort is directed at attracting and retaining all kind of users: farmers, traders, housewives, senior citizens and children. The initial interface is both graphic and voice based to ensure that everyone regardless of their level of literacy can learn to take advantage of the system.

Business Model

The business combines a mother portal, TARAhaat.com, with a network of franchised village Internet centres, or TARAkendras. TARAhaat will deliver education, information, services, and online market opportunities to rural consumers *via* the Internet and its kendra outposts. It also hopes to provide a cost-effective gateway by which larger corporations can reach rural customers. It will offer information, e-mail and web services, and eventually e-commerce and fulfilment services, earning revenues through membership fees and commissions.

Infrastructure

Where possible, TARAhaat will use existing telephone lines to connect its franchise internet centres. But it also plans to install VSAT links where necessary to provide connectivity, utilizing satellites operated by others. Thus TARAhaat is largely dependent on India's existing commercial infrastructure, which is less than ideal. Dependence on the satellites has already caused interruptions in service for some franchises, and is a relatively high-cost solution. Connectivity, and bandwidth capable of supporting a graphics-rich Internet experience, are likely to remain in short supply for the near future, posing a continuing challenge to the venture. In addition, electric power outages are frequent, so much so that the company provides a diesel-powered generator as part of the franchise infrastructure, adding considerably to costs and maintenance needs.

Content

One factor that bodes well for success is the company's flexibility in developing its products. TARAhaat has shown high sensitivity to customer needs, allowing products to evolve to meet these needs. For example, TARAgyan products, TARAhaat's education offerings, came into being only recently to address the strong demand for computer and computer-based education in the initial customer base. In addition, each franchise has latitude in developing products and services that meet local market needs. TARAhaat hopes to create a brand image in which the local TARAkendra is seen as the place for a family to find products aimed at the entertainment, information, and commercial needs of each member. The company provides content in two local languages as well as English, and expects to provide content in other local languages (India has 18 official languages) as well.

Human Capacity

TARAhaat provides extensive support for its franchisees, including assistance with financing, Internet connections, business and IT training, and marketing. It plans to create a TARA university for franchisee training. Many franchises offer computer or other IT classes. Nonetheless, the lack of literacy among many of its intended customers poses continuing challenges to its business strategy.

Policy

India's telecom regulatory policies are improving, but fall short of fully open competition. As a result, prices are relatively high and service is generally quite poor, especially in rural areas. Since TARAhaat is largely dependent on commercially available infrastructure, this poses a substantial challenge to the venture. However, the company faces no major regulatory hurdles.

Enterprise

TARAhaat's association with Development Alternatives brings a wealth of rural expertise and a resilient attitude towards overcoming all hurdles. Some of the initial challenges described in this report have been handled with creative solutions, both at the franchise level and at the enterprise level. However, the venture has substantial financing needs and faces long-term challenges relating to product development and franchise operations that still await resolution. Unlike many startups, however, its management team is seasoned and strong.

Key Lessons

Interviews with users of TARAhaat's services demonstrate the venture's social benefits, including empowering the education of girls, inspiring confidence and higher aspirations among rural children, and enabling farmers to gain market information and substantially higher prices for their crops.

The staying power of TARAhaat's business model is not yet proven, but the positive customer response is an early indication of viability. The venture's franchise model seems likely to harness local entrepreneurial energies to the benefit of both. Its emphasis on locally-relevant content, in local languages, and strong orientation to product development based on customer feedback is also an important characteristic, one that takes advantage of India's large rural middle class. Finally, the company combines both commercial and NGO characteristics, potentially giving it an advantage in pioneering the relatively risky Internet market in rural India, and is unlikely to face major competition in the near future.

Financial Feasibility

The financial feasibility is worked out as follows :

Table 7.4 : Financial Working for *TARAhaat* Initiative

Year	*Costs ($ million)*	*Revenues ($ million)*
1.	0.3	5.9
2.	6.6	15.9
3.	32.1	39.0
4.	72.4	55.5
5.	121.6	80.2
6.	182.0	109.7

The revenues and costs vary from centre to centre. The larger centres achieve profitability in year 1, the medium village achieved profitability in years 3 and the smaller village never achieve sustainable profitability. The small village *kendras* which TARAhaat management considers social investments and not profit generators, will rely on subsidies from TARAhaat which would be offset by profits from larger *kendras*.

Tata Group's Community Initiatives

The Tata Group's commitment to contribute substantially to the community led to the formation of the Tata Council for Community Initiatives (TCCI). With the active participation and contribution of Tata Group companies, TCCI has helped communities to become strong and self-reliant. In another two years time as on 2004, TCCI is concentrating on evolving Tata Corps of Volunteers who will serve as a bridge between the company's explicit intent to develop the community and its employees to provide a broad support. TCCI will strive to evolve simple but business compatible procedures to evaluate the impact of community programmes in terms of their outcomes in human development. The procedure will be titled Tata Social Evaluation, Responsibility and Accountability (Social – ERA).

The M.S. Swaminathan Research Foundation (MSSRF) with the help of the Tatas, has set up the Virtual Academy for Food Security and Rural Prosperity. As a part of this project, the foundation has set up knowledge centres at 10 villages around Pondicherry using information and communication technologies. There is a hybrid wired and wireless network comprising PCs, telephones, VHF, duplex radio devices, spread spectrum, e-mail connectivity, VSAT and a public address system. Not only are electronic media like the internet. cable television, ham radio and loudspeakers used, but traditional methods like yellow pages and community pamphlets and the good old notice board also ensure that the information reaches the intended audience. It is not the hardware that changes the lives of the have-nots. What brings them to the knowledge centre in droves is the data relevant to their lives. MSSRF focuses on helping them convert this know *how* into do *how*.

At Veerampattinam, the fisher folk earlier used meteorological reports provided by State's Fisheries Department. But these sometimes came in too late. The Foundation now downloads the wave height and wind direction from the US Navy website. Its sends an online message in the local language from the main hub to the village centre. The village centre then makes announcements throughout the day on the public address system. Residents of Embalam, among other villages receive not just weather reports and information on the government schemes but also availability of agricultural inputs, market potential and prices of their produce. The project is directed towards meeting the needs of the local community. It works according to the guidelines for sustainable development issued at the 2002 World Summit at Johannesburg. It concentrates on climate management, water energy, health, agriculture and on biodiversity and the ecosystem. The positive impact on women, child education is clearly visible. The people have been helped to develop numerous skills.

Hewlett Packard's Global Citizenship Initiative

Hewlett Packard's (HP's) global citizenship is built around the core of ethics and governance. Fundamentally, the heart of global citizenship is about engaging the communities that the organization works and lives in. The priority areas as a part of their citizenship agenda are three fold: Environment, privacy and e-inclusion. According to HP, the three enablers of citizenship effort are: the government policy, philanthropy and communication. On the environmental front, HP has an explicit goal of designing new products so as to minimize their ecological impact, from production through disposal. The e-inclusion and education programme focuses on using technology to give people access to social and economic opportunities. Regarding privacy rights, HP was the first company in the Fortune 500 to join Safe Harbor, the international data protection agreement outlining rules for transfer of consumer data across international borders.

In their quest for technology solutions for the emerging markets, they developed a solar powered digital photographer's backpack. It takes the standard digital photography equipment – digital cameras, a digital printer which does not require a computer but has lots of intelligence embedded in a kit that runs on solar power. The package is being piloted to stimulate the community in Kuppam by creating jobs for women's self help groups. One of the solutions is an information system for agriculture, a major livelihood in Kuppam, where the women might be employed to take photos of crop problems, The government is also interested in potentially hiring upload on the web and allow local farmers to tap into advice on their crops from experts in other parts of the world. The government is also interested in potentially hiring the women for a broader project to create photo IDs for citizens. Clearly, everyone in Kuppam would not have a digital camera or a photo printer, but individuals can be employed in ways that let them earn a living, using HP's products and stimulate the overall economy.

HP has designated a senior executive with the set of responsibilities associated with the global citizenship programme. When at the HP Executive Council, anything related with HP strategy is discussed, this executive can raise issues about the citizenship agenda. In their quest towards a prototype solution, they initiated an experiment at Kuppam in India. HP has a matrix structure for such initiatives. For example, for the e-inclusion initiative, the e-inclusion market solutions team has a big role. The strategy process has been driven through the creation of horizontal teams made up of key members from each of those departments that has a role in various initiatives.

The Kuppam i-community initiative demonstrates HP's commitment to e-inclusion and community engagement and makes a great deal of business sense. As a high priority growth market for HP, India has many potential customers for its technologies. Kuppam was also chosen because it could help HP learn about nascent or white space opportunities – the high potential opportunities that lie between markets or outside HP's current focus areas. HP feels that their richest opportunities and best chances for innovation lie in regions that rank user and organizational needs very differently from the way established markets and customers do.

When global citizenship efforts and business efforts are being fused, is important to choose initiatives based on the strength of local leadership. The Kuppam I community now encompasses five physical community information centres with more on the areas where students, teachers, parents and others can learn skills and get access to information and services *via* the internet. It features a government portal – usable from a web enabled device which gives access to government's services through the net. The i-community also features a mobile service centre with wireless access and new technology empowered social programme.

Activities of Social Work Research Centre (SWRC)

As discussed in chapter 2, the Social Work Research Centre (SWRC) has been working in rural communities in India to improve the quality of life of the rural poor. SWRC has worked to address basic needs: water, health, education, employment, social awareness and conservation of ecological system. While enrolling individuals in the processes that govern their lives. A voluntary agency, SWRC's main centre is in the village of Tilonia in the Silora Block in the Ajmer district of Rajasthan. The organization began its work in 1972 in Tilonia, by opening a Barefoot College, because a rural development agency could and should not work from a village. Rural development required living among those people who would effect and be affected by that development process. SWRC programmes were initially started with urban expertise from outside the area.

Conclusions

The interventions discussed in this chapter could be summed up as indicated in Table 7.4. The Table 7.4 sums up the initiatives discussed in the chapter on various parameters. These parameters are: objectives of the intervention, stage of development, geographical reach of the intervention, beneficiaries of the intervention, financial feasibility, dependence on grants, ability to sustain the operations of the firm, range of intervention, impact areas of intervention; networking using the intervention. These variables can be the yardstick for making comparison across various interventions.

Table 7.5: Interventions and their Context

Initiative	*Objective*	*Stage of development*	*Geographical area*	*Beneficiaries*	*Financial Feasibility*	*Dependence on grants*	*Self - sustaining*	*Range of Intervention*	*Areas where impact is created*	*Network with others*
ITC	Supply chain Management	Medium	Central and north India	Rural farmers	Yes	None	Yes	Narrow	–Distribution network –Information access –Community exposed to IT networks	Yes Primarily with community, fertilizer firms, government and other business organizations
TARAhaat	Information products and services through Internet	Varies from region to region	Widespread	Rural Markets	Yes for more developed areas. No for less developed areas.	Yes for under-developed areas, primarily managed by equity investments from the corporate sector	Yes in the long run	Narrow	–Village connectivity –access to products and services needed by the farmers	Employment Empower-ment Delivery of products Decentralised university.
M.S. Swaminathan	Corporate intervention along with an NGO Used IT for generating food security and rural prosperity	Low	South India	Rural communities	N.A.	Yes	Still evolving	Narrow	Climate management Water, energy, health, agriculture Biotechno-logy	NGO and corporate net working

(*Contd.*)

Table 7.5 (*Contd.*)

Initiative	*Objective*	*Stage of development*	*Geographical area*	*Beneficiaries*	*Financial Feasibility*	*Dependence on grants*	*Self-sustaining*	*Range of Intervention*	*Areas where impact is created*	*Network with others*
HP's Global Citizenship	Engaging communities at grassroot levels for product development	Medium	South India	Rural communities	Yes	No	Yes	Narrow	IT usage for reemploy-ment Women empower-ment	Corporate and community networks
SWRC (Barefoot College)	Wide range of Interventions targeted at extreme poverty conditions	Low	Initiated in North West and then shifted to different parts of India and other countries as well	Rural areas	Yes in the long run if coupled with corporate initiatives	Yes	Yes	Wide	Education, Agriculture, Food, Wasteland development, Solar energy, Women empower-ment, Shelter, Water	Networking of community, government, NGO and international agencies

There is no unique solution for development efforts. Different geographical areas are at a different stages of development and, therefore, one solution cannot be an answer. A mix of interventions is required. The solution may have technology which is very basic or a technology which is high tech. Also for implementing such solutions, there is a need for interaction between the community, corporate firm, NGO and the government. One solution cannot fit all geographical area.

Intervention such as Barefoot College

Barefoot College intervention has been initiated in an area where there was little by means of livelihood and sustenance. There was nothing by way of education, healthcare, basic infrastructure and drinking water. In such a scenario-installing a computer would not solve a large number of issues. The basic needs of people need to be fulfilled before high tech interventions are initiated :

1. In such a scenario, very basic technologies would have to be used.
2. The interventions were initiated by an NGO which raised funds from international organisations.
3. It is a remarkable case of community led intervention with a wide scope of activities. Initiatives which first lead to have access to basic amenities have an impact on the self esteem of individuals. Then other interventions can be more successful.
4. It did not involve any experts but individuals who felt that the community would benefit if they themselves identified with the place and the people with whom they worked. Participation and identification with community is a prerequisite for success for any intervention. External consultants cannot be a long term solution for any endeavour. It is, therefore, essential that local community itself gets involved in their own development. The leadership role must also be taken by the local stakeholders so that there is a higher degree of seriousness.
5. At a later stage of development initiatives such as those undertaken by HP or ITC could bring it to the next stage of development.

Intervention such as HP Global Citizenship Programme or TARAhaat

Interventions such as these must be initiated in a region where basic amenities are available. When people have access to basic infrastructure have an access to education and food, then a means for sustained livelihood becomes important. Herein the HP initiative could make a major difference. But such an intervention where nothing is available, may not make that much of a difference:

1. Initiatives such as initiated by HP are directed at a community who have some resource base for survival. Here there are people who have some means for living and have some level of literacy and access to food and water.
2. An initiative such as the one initiated by HP makes sense as it provides employment and empowerment for the communities.
3. It helps the firm develop and align products which could not only be directed at urban population but also at the grassroots level.

ITC Initiative

This intervention is unique as it blends the corporate imperative for profitability with the resource base of an area. It leads to a win-win situation both for the community and the corporate firm. It leads to a scenario wherein the community benefits from corporate intervention:

1. The initiative such as the one initiated by ITC in India is an interesting case. There is a resource base available.
2. The effort is directed at system inefficiencies and then linked with improving procurement from the villages. It results in cost savings by the farmer.
3. It opens new doors of opportunity for the firm for additional revenues. It gives it impetus for expanding into more areas for procurement. ITC started with soya procurement and then moved over to wheat procurement.
4. Here initiatives such as those by HP and Barefoot could bring in a holistic development and empowerment of communities.

The interventions that are initiated have to be sustainable and self propagating. So an important dimension is the financial feasibility to create a spiral effect that leads to that region's growth. The ITC's model is an indication that being commercially savvy and also have community orientation is possible. There is a huge potential to convert the latent demand into an active demand and bring in new numbers in the list of potential consumers.

The involvement of the local community and their role in participating makes a huge difference. There is a need for redefining the roles and also make them understand the outcome of such interventions. When there is a realization that such interventions will lead to a larger good of the community then their motivation levels are higher. Corporates would have to look for such unique value drivers which may not just come through artificial means such as expenditure on advertising. After all lot of the products that actually sell in the market are because of the superficial differentiation. So efforts which are directed at doing things differently which lead to a positive impact on the resource usage and recycling the existing resources would be desirable.

ITC management of the bureaucracy is also unique. When e-chaupals were conceived, they faced a fundamental regulatory obstacle. The agricultural Produce Marketing Act, under whose aegis *mandis* were established, prohibits procurements outside the *mandi*. ITC did not use the *mandi* infrastructure and invested on building the electronic infrastructure. Since ITC paid the tax on the produce sourced, it did not risk the relationship with the government and *mandis*. So solutions have to be generated which empowers people and involve them in the decision making process. Interventions targeted at any area would have to be matched with the needs of the area. Also the effect must be simultaneous, i.e., water, shelter, employment, health care, literacy will all have to go hand in hand to make the intervention sustainable. The network of a number of organisations will have to be created to achieve this goal.

NOTES

1. The World Bank (2003) report indicates that in the year 2001, 72 per cent of the people residing in India were the rural population (p.121). The population density was 454 (people per sq. km of arable land) in the year 2000. In India, in 2000, the land area was 2,973 sq km. Out of this arable

land was 54.4 per cent. Permanent cropland was 2.7 per cent of the land area. The use of agricultural machinery is low in India as compared to the developed countries. In India, the number of tractors per 1000 agricultural workers were 6 in 1998-2000 as against 2 in 1979-81. The number of tractors per 100 sq. km of arable land in 1998-2000 were 93 against 24 in 1979-81. Compared this with the figure in Israel where it stood at 734 per sq km or 4,691 in Japan, 1,549 in Norway, 2,512 in Austria. Poor farming practices can cause soil erosion and loss of soil fertility. Productivity improvements through excessive use of chemical fertilizers can lead to pesticide poisoning and alter the chemistry of soil.

2. There have been sporadic corporate experiments which have resulted in success for some firms.These have been tailored to suit rural requirements Arvind Mills in India found that even the cheapest jeans for most Indians were out of reach. It introduced the ready to stitch Ruf & Tuf jeans kit, priced at Rs. 195 against the organized sector price of Rs. 300. The product was available in local retail outlets serving the hamlets with populations ranging upto 5,000. Village tailors were trained and provided with additional machine accessories needed for stitching heavy Denim. The result was that the five million kits were sold within 18 months. Products need to reach the villages with over 2,000 people (there are about 65,000 of these) accounting for 40 per cent of the rural population and 60 per cent of the rural consumption.

3. Terms used in the text

Choupal	A Hindi word which connotes a village gathering place
1 crore	Indian term for the number 1,00,00,000
Lacs	Indian term for the number 1,00,000
Kendras	Centres
Madhya Pradesh	An Indian State in central India
Mandi	A government mandated market place where farmers sell their crops or other things are bought and sold.
Rajasthan	An Indian State in North West India
Quintal	One hundred kilos
Rupee	Indian currency
Samyojak	Commission agent in his role as a collaborator
Sanchalak	Village farmer who runs the e-choupal and acts as ITC's representative in the State
VSAT	Satellite based technology which helps in wireless connectivity.

REFERENCES

Annamalai, K. & Rao, Sachin (2003), *What Works: ITC's e-Choupal and Profitable Rural Transformation*, World Resources Institute.

Assisi, C. & Gupta, I. (2003), ITC's Rural Symphony, *Business World*, January 10, India.

Choi, S.Y. and A.B. Whinston (2000), *The Internet Economy: Technology and Practice*, Austin, Texas: Smartcon.com.

Choi, S.Y. *et al.* (1997), *The Electronics of Electronic Commerce*, Indianapolis, Indiana: Macmillan.

Dunn, D. & Yamashita, K. (2003), *Microcapitalism and the Megacorporation*, Harvard Business Review, August.

Grenier, Louise (1998), *Working with Indigenous Knowledge*, A Guide for Researchers, IRDC, 1998.

Holliday, C. (2001), *Sustainable Growth, the DuPont Way*, Harvard Business Review, pp.129-134.

HP. com (2004), *Global Citizenship*, Retrieved on July 12, 2004 from http://www.hp.com/cgi-bin/hpinfo.

Human Development Report (2003), Oxford University Press, New Delhi.

Levine, J. (2004), The citizenship agenda at Hewlett Packard: An interview with Debra Dunn, Knowledge Centre, Retrieved from (http://www.be.edu/centres/ccc/pages/resources/kn_resart_dunn.html).

Meghani, S. (2004), *A Village Comes of Age*, http://www.tata.com Retrieved on August 27, 2004.

Peterson, C.; Sandell, V. & Lawlor, A. (2001), *What Works: TARAhaat's Portal for Rural India*, World Resources Institute, July.

Porter, M. (2003), *Building the Microeconomic Foundations of Prosperity: Findings from the Microeconomic Competitiveness Index*, The Global Competitiveness Report 2002-2003, Oxford University Press, New York.

Reinhardt, F. (2003) *Tests for Sustainability, The Global Competitiveness Report 2002-2003*, Oxford University Press, New York.

Roy, Bunker (2001), Management of Water Resources at Local Level—The Repair and Maintenance of Hand Pumps, http://www.unv.org/projects/sl/collect/swrc1.htm.

Sarvani, V. (2003), *ITC's e-choupal: Taking e-business to farmers*, Case Folio, July. Also available from www.icfaipress.org.

Sharma, M. (2000), *Information Technology for Poverty Reduction*, Retrieved from *www.ouhk.edu.hk/cridal/wrapup/discuss/messages/22.html.*

TARAhaat (2004), *What Works: TARAhaat's Portal for Rural India*, Retrieved from http://www.digitaldividend.org/case/case_TARAhaat.htm Retrieved on Sept. 1, 2004.

Transparency International (1996), Sharpening the Response against Global Corruption: *Transparency International Global Report* 1996, Berlin.

Turban, E.; King, D.; Lee, J.; Warkentin & Chung, M.H. (2002) Electronic Commerce, New Jersey, Prentice Hall.

World Bank (2003), World Development Indicators, The World Bank, Washington.

8

Towards A New Framework for Sustainable Development

Based on the studies in chapter 7, a framework is being proposed which links various geographical areas on the basis of their stage of development with the nature of intervention which may work.

Different geographical areas are at different stages of evolution. They differ in infrastructure – literacy levels, availability of basic amenities, employment opportunities, environmental conditions, presence of industry. With all these factors, it is important that there is a match between the intervention and the State of development. We could take technology as one of the tools which could facilitate development. It alone may not be sufficient tool to determine the complete intervention. For an intervention to be complete, it would need softer components and some technology as well. Consider an area which has low levels of literacy, limited resource base, dependence on basic produce, lack of basic amenities of life. Such an area would need interventions targeted at the basic needs to which people do not have an access. Such efforts would require the efforts of an NGO, local community, international organisations as well. Such an intervention must be broad based in its approach. The other aspect is the use of sophisticated technologies which could be tapered to the needs of the local community. The most important aspect is the matching of the environmental strengths and needs of the local community with products which result in win-win situation for both.

In India, there are many strengths. India is one of the world's largest provider of fruits and vegetables while our share in the global food industry is negligible. India has a large resource base. So interventions that create larger markets, makes these communities economically independent, enabling them to have a sense of self- esteem.

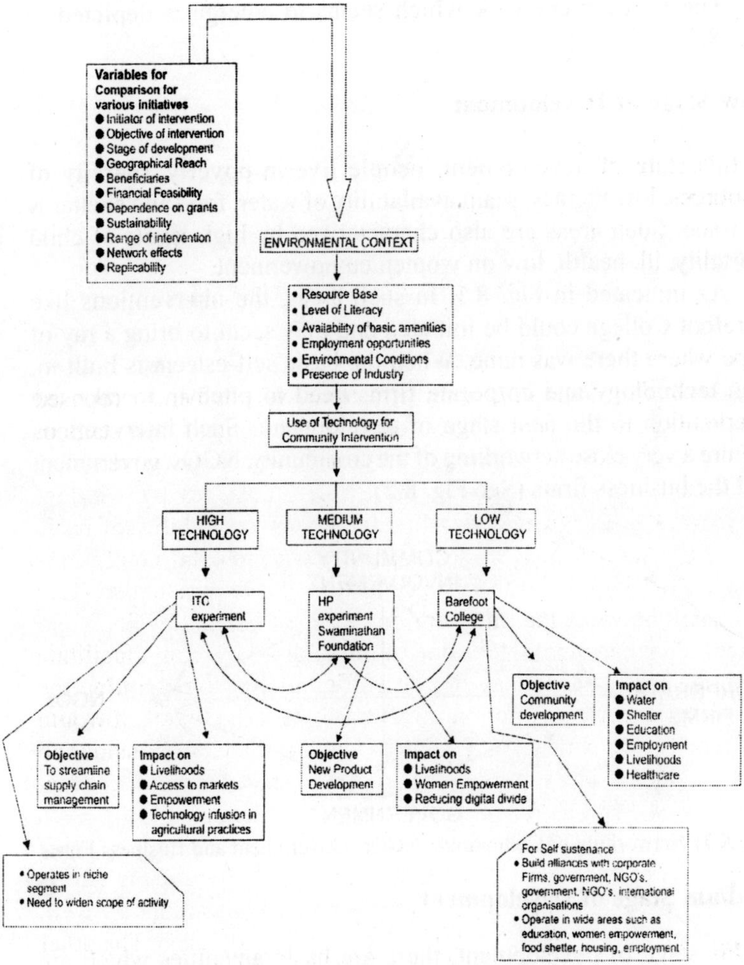

Fig. 8.1: Alternate Models for Sustainable Development

The broad framework which seems to emerge is depicted in Fig. 8.3.

Low Stage of Development

In this state of development, people live in poverty, scarcity of resources, low literacy, scant availability of water, food and shelter is an issue. Such areas are also characterized by high levels of child mortality, ill–health, low on women empowerment.

As indicated in Fig. 8.1, in such areas, the interventions like Barefoot College could be initiated and they seem to bring a ray of hope where there was none. When the basic self-esteem is built in, then technology and corporate firms need to pitch in to take the intervention to the next stage of development. Such interventions require a very close networking of the community, NGOs, government and the business firms (See Fig. 8.2).

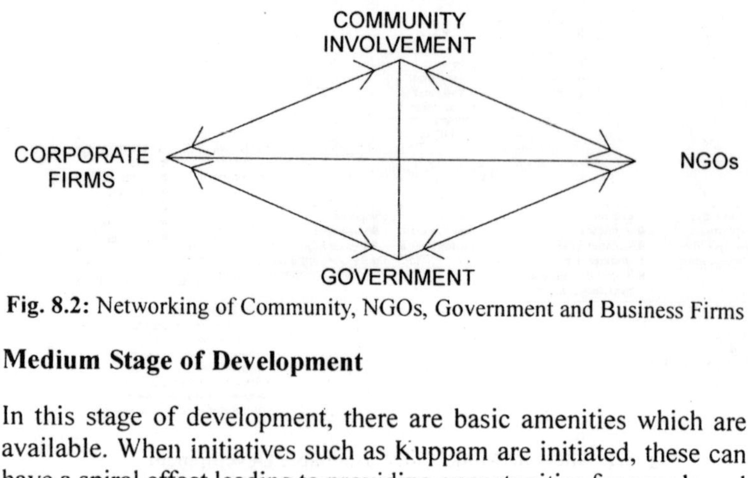

Fig. 8.2: Networking of Community, NGOs, Government and Business Firms

Medium Stage of Development

In this stage of development, there are basic amenities which are available. When initiatives such as Kuppam are initiated, these can have a spiral effect leading to providing opportunities for people and the corporate firms as well.

Higher Stage of Development

The next stage is where there is resource base, economic activity is happening but there still are areas where there is an exploitation. By looking at areas which bring in efficiency into the system by making supply chain interventions as done by ITC can help the community leapfrog into the next stage of development (see Fig. 8.3). This can have an immense consequence. At this time if support is provided by NGOs and initiatives comparable to Barefoot initiative are undertaken, they could lead to a holistic development (see Fig. 8.3).

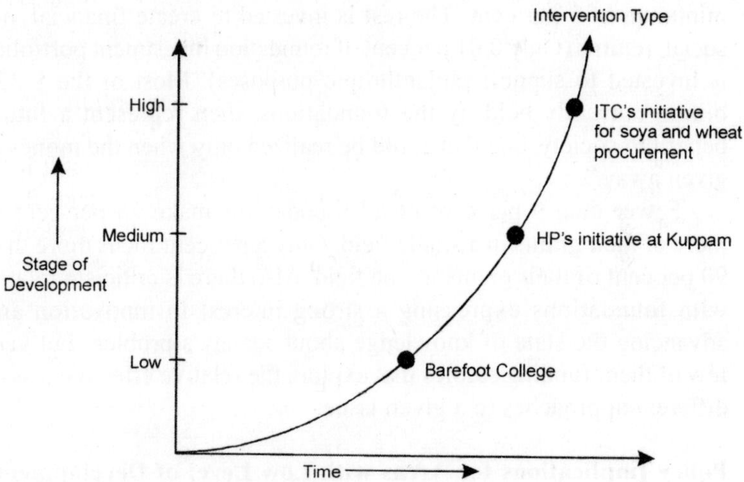

Fig. 8.3 : Types of Interventions

Critical Factors for Success

Synchronisation

All these interventions need to be in synchronization and various stakeholders have to work together. A lack of support from anyone of them could hamper the overall development process.

Use of Local Resources

There are issues about access to information and actually initiation of efforts. There are many institutions, NGOs, trusts and voluntary organizations which initiate numerous efforts. There needs to be documentation effort on the part of the government wherein all the contact points and area where work is carried out is clearly indicated. Consultants and experts need to be kept out of such interventions. A study reported in Harvard Business Review points out at some alarming facts about the intervention initiatives. There is a pointer towards the foundations.

"When a donor gives money to a foundation, most of the gifts sit on the sidelines. On average, foundations donate only 5.5 per cent of their assets to charity each year, a number slightly above the legal minimum of 5 per cent. The rest is invested to create financial, not social, returns (Only 0.01 per cent of foundation investment portfolios is invested to support philanthropic purposes). Most of the $ 330 billion currently held by the foundations, then represent a future benefit to society, one that could be realized only when the money is given away."

Fewer than 9 per cent of all foundations make 75 per cent or more of their grants in a single field. Only 5 per cent focus more than 90 per cent of their grants in one field. Also there is criticism related with foundations expressing a strong interest in innovation and advancing the state of knowledge about society's problem but very few of them funding studies that explore the relative effectiveness of different approaches to a given issue.

Policy Implications for Areas with Low Level of Development

So for areas where resources are limited some of the following interventions are suggested:

Simple Technologies

Infusion of capital or technology alone is not a solution to the problem of poverty. It is a multifaceted issue and requires interventions at various levels. Merely putting in computers in the villages is not a solution. The reality is that there are large number of people who are illiterate, are unemployed and need to be employed meaningfully so that they could earn their livelihoods. In the cases seen above in various sectors it has been observed that the technologies which were used were basic, were easy to understand and people have been knowing these for long times. So the solutions were not imposed from outside but were generated by the communities themselves. The case of the Barefoot College at Tilonia indicates that even the illiterate people can be trained if the technology is easy to understand. Simpler the technology is, higher would be its acceptability by the people.

Documentation of Indigenous Technologies

This points to the understanding of the importance of the indigenous knowledge. Despite technology advances, the problem of drinking water and sanitation is still prevalent. In the olden days, they did not have trained doctors, architects or biologists. They did not depend on theory but applied the wisdom they had acquired over the centuries. Many communities are so fed up with the current state of affairs that they demand they be left alone to identify and solve their own problems (Roy,1999). The indigenous knowledge helps to arrive at solutions. Greiner (1998) defines indigenous knowledge (IK) as "the unique, traditional, local knowledge existing within and developed around the specific conditions of women and men indigenous to a particular geographic area". Indigenous knowledge can provide insights into the area of food security, health, education, natural ecological initiatives. The advantage of the indigenous knowledge is that it is dynamic in character, as it has the capacity to evolve since people have been using them in the past for surviving through centuries. It needs documentation, stored in systematic manner as it has been passed on from generation to generation. The solution does not lie in inflicting western solutions on communities. Interventions fail to induce people to participate because of the absence of instruments and mechanisms that enable them to use their own knowledge.

The use of traditional knowledge, skills and wisdom promotes active community involvement because people depend more on each other. The use of traditional knowledge demystifies the local technologies that will be the basis for sustainable solutions in the future. The more people who understand and try out a technology, the greater the chance of the technology getting accepted. In 1997-98, through the use of centuries old technologies, a total of 12 million litres of rainwater was collected in 100 schools attended by 3,000 children at the Barefoot Schools. The cost was a mere US$ 0.10 a litre. The schools have teachers with no qualifications. Over 150 young people from nine States of India have been trained as barefoot solar engineers. They have equipped over 2,000 houses in the Himalayas with solar electricity. The practice could be transferred to other places and situations, but it is essential that several conditions be met (www.unesco.org/most/bpik16.htm), The Barefoot College – Promoting Productive Employment for Youth.

The characteristics of these solutions have to be :

1. The technology solutions have to be rooted in the ground realities,
2. They have to be sustainable,
3. They have to accommodate huge numbers, and
4. They have to trap investments.

Solutions to be Built around the People through their Involvement

The people need to be kept in mind for whom the solutions are being generated. Another important aspect that is important is the total involvement of the people at the grassroot level. No external agent can bring about the change in the community externally. Unless the people stand up for themselves, nothing can change. They would have to be united together to stand up for their rights and there is a need for facilitators rather than consultants. These should be the people who can identify themselves with these people and live with them so as to experience what it means to live a life these people live. The solution for any community has to be a sustainable solution that leads to a process, which is self-sustaining. It can only happen if the use of the local know-how is usefully channelised so that it can be sustained in the long run. The case of Tilonia as discussed in the chapter is an example of the same. The people need to be sensitized to the change. Acceptance of change is far easier when there is a suitable climate created for the same.

Dissemination of Information to the Poor

The biggest problem of the poor people is the lack of information on what is happening around them. The government starts so many programmes for the poor people but the benefits never reach down to the poor people. The government needs to communicate the information to the poor about the programmes and most important monitor the implementation of the same. For example, the much hyped Swarna Jayanti Sahari Yojana and Swarn Jayanti Swarojgar Yojana, which was launched in April 1999, have failed to benefit even 5 per cent of the target in Orissa. Bamboo craftspersons in Bhubaneshwar which were supposed to be the beneficiaries of the Swarna Jayanti Sahari Yojana, was meant to help people rise above the poverty line.

In the last two years, since the scheme began, not a single person received even a rupee as a loan (www.ndtv.com, Mar. 1, 2001), Pro- poor Policies let down by the lack of reform).

Better Management of Resources

A recent news reported in the press is an eye opener. The drought affected people in ten districts of Rajasthan protested to highlight that though the Food Corporation of India godowns were bursting with food grains, people remain hungry. In godowns in India, the FCI has nearly 410 lakh tonnes of foodgrain, nearly an estimated 139 lakh tones is in excess. It costs the government Rs. 4,200.00 lakhs just to store these foodgrains in Rajasthan (www.ndtv.com, May 17, 2001). The Food Corporation of India has in recent years been grappling with the problem of how to take care of the huge surplus foodgrain which is fattening ever growing army of rodents or simply rotting away in its badly managed warehouses. (*Times of India*, May 11, 2001). So there is a need for a better management of the resources. With little changes in the way things are operated, it becomes quite easy.

Political Will to take Pragmatic Decisions

The income and consumption patterns of the poor are changing. The shift is away from course grains to wheat, paddy and oilseeds. Singh (2001) has analyzed the dynamics of the cropping patterns. A drop of 50 per cent in the cultivated area of sorghum, little millet and finger millet has come about just in the past decade. It was in the 1980s that the Public Distribution System (PDS) became a welfare instrument to provide essential items at nearly half the market price. Neither crop loans nor crop insurance are available for these groups. Also there are no subsidies. The promised minimum support price of coarse grains is denied to farmers due to government non-intervention. The chemical composition of coarse grains is better than rice and wheat in many cases. Pearl millets have a higher concentration of protein, fat and minerals particularly calcium.

The 1999-2000 Union Budget projected a cut in the central deficit of 0.9 per cent of GDP. Achieving this target depends on substantial rise in tax revenue and containing revenue expenditure growth to only 9 per cent. The interest costs of the debt have increasingly crowded out infrastructure, maintenance and social spending in central and State budgets. Implicit and explicit subsidies at the centre and especially at the State levels are a major factor in the deficit (World Bank Report, 2000). The Ministry of Finance estimated these subsidies at over 14 per cent of GDP in 1994-95. In addition to increasing the deficit, they are distortionary, non transparent and at best have uncertain equity consequences. At worst they are anti equity.

Transparency

Transparency in operations is required. A review of literature suggests that poverty has a relationship with political corruption. Oyen (1997) in Human Development Report (1997) pp. 95 mentions that poverty often serves the vested interests of the economically powerful, who may depend on the poverty stricken to ensure that their societies run smoothly. A mobile pool of low paid and unorganized workers is useful for doing the "dirty, dangerous and difficult" work that others refuse to do. Corruption in government increases poverty in many ways. Most directly, it diverts resources to the rich people who can afford to pay bribes and diverts resources away from the poor people who cannot (Transparency International, 1996). Corruption also skews decisions in favour of capital-intensive enterprise rather than labour-intensive activities which are more likely to benefit the poor. Corruption also weakens the government and lessens their ability to fight poverty. It reduces tax revenues and thus the sources for public services. Most generally, corruption eats away at the fabric of public life – leading to increased lawlessness and undermining social and political stability.

Where corruption is rampant and evidence is in place that development funds have never reached the target segment, it is all the more required. The Barefoot College is evidence where through the *Jansunwai*, it shared all the financial details with the people for whom the development work is being done. It must involve participation by the poor if it were to yield meaningful inputs. It is necessary to vitalize the community media with the involvement of people (Sharma, 2000). It is important to bring information to the doorsteps of the poor people. There is an urgent need to ensure access to modern information technology in rural areas or disadvantaged communities to disseminate simple, practical knowledge which will save their lives, increase awareness and stimulate development. Properly used, media can help reduce the conflict and strengthen local organization. It can help reduce poverty through providing information on how people of their type somewhere else are handling their situation. The poor have inadequate access to information, technology, expertise and resources.

Linkages between Different Sectors

Interventions taken in one sector have an impact on the other sector and there are strong inter-linkages between the same. Improving health outcomes not only improves well-being but also increases income-earning potential. Increasing education not only improves well-being but also leads to better health outcomes and to higher incomes. Providing protection for poor people not only makes them feel less vulnerable—it also allows them to take advantage of higher risk, higher return opportunities. Increasing poor peoples voice and participation not only addresses their sense of exclusion—it also leads to better targeting of health and education services to their needs.

Survival within poverty includes many strategies, which are combined in a process aimed not just at income in the broadest sense but also at assurance against the stresses and shocks to which poor people are particularly vulnerable. One such strategy is a mixture of jobs, some temporary, some full time, some self employed, some working for others. Technology could play a vital role in the elimination of illiteracy. For example, the information technology could be utilized to make the education reach to the most distant location. It could create a difference by dissemination of information using various means in the rural areas. It could address concerns like addressing the problems faced by craftsmen, a farmer or a person who runs a leather tannery among others. Another important aspect that can be addressed by the IT could be converting entire literature in a language, which the masses understand. But IT alone cannot be a solution. The solution has to be rooted into the reality of the situation.

Areas where Resource Base is Stronger

In such areas it is important, that technology is incorporated which improves the productivity. Some interesting conclusions can be drawn from ITC's experiment. These are :

Have a Clear Focus

Business firms need to have a clear sense of direction. Risk analysis is important. From ITC's perspective, the risk was high if it were to continue with tobacco business. By diversifying into areas where India is resource rich and by imbibing requisite technology, ITC ventures into an area which can fetch good returns for the firm and also bring in tremendous goodwill. Profits are important to add value to the investments made in firms. They have their own role. They generate employment and bring in development as well. But such initiatives have inclusion effects and also expand the potential market of the firm.

Identification of the Customer Needs

Need identification is important. Herein customers need branded products which assures them of quality and value for money. However, the organization at any point of time is always trapped in tradeoffs between profits through market research and investments in innovation and societal benefits. It is not just advertising expenses but also other value drivers that could contribute towards organizational excellence. In ITC's case the entire product line may shift as it has in its attempt to streamline the supply chain, it has not only made the procurement process more effective, but has made a tremendous contribution to the society as well. It has improved the wages that the farmer carried home for his hard work. This assures a better quality of life for him. Also it meant replicating the experiment for not only soya crop but also extended it to the wheat as well. So this experiment takes the entire concept of branding and development to a new paradigm which is not just caring for profits but developing products that ensures sustainability.

Redefine Roles

The idea is not to take away from people but modify the roles and responsibilities so that they have a sense of self esteem and feel involved. Such efforts can be successful if the community has a feeling of trust and feels that it will benefit out of the process. So community involvement and redefining the roles are important.

Networking of Different Stakeholders

At some point different stakeholders would have to work in synchronization with each other. Each intervention will have a range of activities and will be constrained by the state of development. The initiatives for the overall development must include simultaneous interventions by many firms, government and NGOs. More than institutions, individuals would have to come forward and take a lead in such initiatives. The bigger organizations have lot of administrative costs. The benefits need to trickle down to the lowest level for a real change to happen which is self sustaining in character.

9

Conclusions and Recommendations

An Overview

While researching for this book we came across numerous case studies, which ultimately pointed towards two very pertinent questions: (a) Despite development in both industrial and agricultural sector why has poverty continued to plague the Indian masses? (b) Why has India failed to increase the per capita income of the rural poor? And we realized that we were in total agreement with Little, who in 1990 stated that its (India's) failures were daunting; it had made little or no progress in reducing the mass and depth of poverty since Independence, and inequalities—particularly rural inequalities—remain high by Asian standards. In spite of a national development rhetoric that gives priority to poverty alleviation, the condition of the poor (largely rural) has changed very little after the independence.

In our quest for answer we came across Atul Kohli's *The State and Poverty in India.* In his book he states that "the welfare of the poor is not correlated with the overall prosperity of a state. Instead, the critical variable is the type of regime in power during the process of economic development. Regimes formed by strong, competent political parties of the left succeed in tilting the process of development towards poverty alleviation, whereas weak regimes and regimes dominated by the propertied classes have a poor record of performance in poverty reform. Poverty reform requires a political regime that has both the will and the means to implement it, and a regime that is relatively autonomous from the political reach of economic elites". Besides, economic development processes also result from (a) government economic policy, (b) the private activities of national and multinational corporations, (c) the influence of industrialized-nation governments, and (d) a variety of bilateral and multilateral development agencies. None of these agencies work in isolation and it has been proven that interventions in one sector often has impact in the other sectors as well.

Financial Interventions

Poor need to be financially sustained at initial stages of any entrepreneurial effort. Principally, these communities either exist in the rural India or at the fringes of the urban society. Comprising of mostly women, these communities are considered non-bankable unworthy of any credit. Mostly working as daily wage earners these workers have for long depended upon the capricious nature of the wages handed down to them by the labour contractors. Streamlining and fixing of minimum daily wage slabs have altered the earning and spending capacity of this strata. Numerous micro-finance schemes and Grameen banks have thus been set up to promote the off take of credit on small scale. It has encouraged them to form self-help groups which collect small amounts of money for subsequent disbursement of small loans among its members, thus developing mechanisms to provide the poor with greater access to credit. This has facilitated induction of simple technologies and access to markets. Many small scale enterprises like retailing vegetables, photography, weaving, video film making, etc. have been set up with the credit thus available. Children and young adults of such households now have access to better schooling and other infrastructural facilities like markets, water, adequate shelter and opportunities of better employment. Financially, a large part of the poor are becoming less vulnerable and more empowered to take advantage of higher returns of prudently managed finances.

Educational Interventions

Even though an infrastructure for the schools have been established with the help of the local community, it may not be able to deliver the results for lack of teachers. Often due to remote locations of the villages and because of lack of any connectivity to the nearest town, teachers are reluctant to work in such schools. Moreover, problems of family, children and the societal attitude act as a barrier for the lady teacher to teach in village schools. Teaching becomes irregular in such circumstances and this impacts the student attendance and maintenance of school resources like the building, toilets, blackboards and the other teaching material. The issue of schooling and skills, therefore, takes on a new significance because of the growth of protected labour markets, particularly for employment in international, manufacturing and clerical jobs, in newly-created employment/industrial infrastructure. "In these ever changing/expanding labour markets, schooling and the credentials it provides are used as selective mechanisms and as a measure of labour-trainability and discipline. Access to high wage jobs is not only dependent on being competent in certain skills but also related to the protection and bargaining power of organized labour"(Heyzer, 1992), Interventions such as holistic literacy programme of the Barefoot College of Tilonia, therefore, demand a special merit because they exhibit total community involvement, consensus building, preparation, sensitization and development of local human resources for indigenous progress.

Health and Hygiene Interventions

According to *The World Health Report*-2004, "The actual capacity of communities to participate in defining and implementing health agendas has been limited by resource constraints, entrenched professional and social hierarchies, and public health models focused on individual behaviours and curative biomedical interventions. Gender, race and class discrimination also play a role. Nevertheless, communities have taken part in many successful public health projects, including sanitation, nutrition, vaccination and disease control programmes. Recent reviews of primary health care have continued to find strong support for community participation and there is evidence that such involvement has led to 'significant health gains'. As has been exhibited by the Sulabh International, any community commitment that is devoid of issues regarding employment/employability, education, and training is reflective of a symptomatic approach towards empowerment. A composite health and sanitation initiative like that of the Sulabh International is indicative of a three dimensional approach, viz., Information, Intervention and Innovation towards poverty alleviation and sustainable development.

Interventions in Watershed Development

A country with all the major geographical zones, India has had a challenging task to manage its water resources. However, over the years, rapidly declining fresh water sources have had a dramatic impact on sustainable development. This can be observed in increased migration of cattle and human population, degradation of arable land, landslides and choking of natural aquifers. Not surprisingly water and environmental development have become a priority area both for the government and the citizens.

As a crucial necessity, water scarcity has affected agriculture, industry, and lives of ordinary citizens with identical intensity. Efforts such as rainwater harvesting, reforestation, restoration of the dried aquifers, desalination and pollution control, and management of flood waters, by the local communities and NGOs, have at best been peace-meal. It is imperative that the paradigm of maintenance, restoration, recharging, and recycling be religiously followed. These initiatives need to be interconnected so that a semblance of balance can be maintained by distributing the surplus into the deficit areas.

Intervention in Empowerment

As the earning demographics change, the fundamental concern in empowerment strategies begin to reflect a growing need to address those factors that constrain the capacity of poor people, along with those that encourage them to pursue their aspirations, to exercise their rights and to engage in gainful productive activities. The changing economic profile of the country has brought to the forefront the role of women as the bread earners. For many of them household chores are no longer unprofitable, yet in the absence of any vocational talent, other than routine household activity, these women feel constrained. Sri Mahila Griha Udyog (Lizzat Papad) has tapped this immense potential to create an enterprise unlike any other in the world. SEWA has taken this concept a step further, its members have taken upon themselves to establish the women in the community as a financial force. Both of them have been instrumental in endowing the women with greater assets in the form of property, tools and employment avenues. This has emboldened them to assert themselves in their household, in decisions relating to the education of their children and to their health. By taking the proverbial 'bull by the horns' these women have forced a rethink on commercial sustainability of various initiatives regarding women empowerment on a world platform.

Interventions in Livelihood

Bringing small dairies of the entire village under the aegis of a single cooperative is perhaps the single largest step in the direction of poverty alleviation. It virtually "set the cat among the pigeons" by modernizing the mind set of rural poor. It compelled them to consider a scientific way of life which, also made economic sense.

Cooperatives have revolutionized the way business is done in rural India, not only have they opened up avenues for rural enterprise, they have also empowered the marginalized sections of the society to enter the mainstream. The chief function of these cooperatives, besides coordinating production has been to alter the economic environment within which dairy farming took place and to discourage the re-emergence of exploitative forces in the countryside. Introduction to modern management techniques, accountability, financial transparency in day-to-day operations have made these enterprises competitive in the international business. Cooperatives such as AMUL has established a benchmark for the farmers, and governments across the world to emulate. Cooperatives have unleashed a tidal wave of economically sustainable development around the world, affects of which are still being felt the world over.

It can thus be stated that at present India is being subjected to a whole range of anti-poverty and pro-development initiatives, targeted at the poor. A few of them have set a benchmark for the world to follow, others have made remarkable inroads into sustainable developmental efforts. As students of this process we feel that more still needs to be done, especially through citizen participation and corporate contribution, because we can no longer depend solely on governmental support or afford to look the other way.

Our country is at present poised on the verge of an explosive economic development, we cannot, therefore, turn a Nelson's eye to the vast poverty stricken diaspora of this country, whose degraded condition will not augur well for any pre-eminent position that this country is likely to achieve. However, as we have reiterated throughout this endeavour, we strongly feel that no development can be considered comprehensive if it is devoid of humanity, sensitivity and empathy towards the less fortunate.

Women Empowerment : Some Recommendations

The concept of women's empowerment implicitly assumes that in all societies, men control women—or, to be more precise, men control at least some of the women of their social class, particularly those in their households and families (Mason and Smith, 2003). Hence, the realization of women empowerment is dependent on the recognition of rights of women, both at home and in the society. It is important for development because empowerment determines the extent to which women gain access to education, are able to seek employment or health care outside of the family, can acquire contraceptive information, and have the freedom to act on their fertility preferences or on the illnesses of their children, among other dimensions (Caldwell, 1986; Dyson and Moore, 1983; World Bank, 2001). In this regard the persistence and leadership of various agencies cannot be underestimated. Still, the lacuna of comprehensive empowerment strategy for women is glaring due to the absence of a pan Indian programme. This brings to mind few lines from Atul Kohli's book *The State and Poverty in India,* wherein he states that "the welfare of the poor (read women) is not correlated with the overall prosperity of a state....Poverty reform requires a political regime that has both the will and the means to implement it, and a regime that is relatively autonomous from the political reach of economic elites".

It is our understanding that "Women more readily embrace the collaborative perspective needed to cut through ethnic, religious, tribal and political barriers. They need support, however, to translate their pragmatic expertise into participation in formal peace negotiations, in peace building and reconstruction, and eventually in national governments. Decisions that exclude half the population from political processes offer little hope of achieving a secure and peaceful future free from violence and discrimination" (Heyzer, 2002/2003). While researching for this chapter, we were thus able to identify four major thrust areas, which could act as guiding posts for developing structures for women empowerment in: The Government, The Corporate and the NGOs, The Community and lastly, The Women themselves. It is our endeavour to identify functions/duties/areas of intervention, so that a holistic drive can focus on economic and social justice which makes empowerment possible.

What should be the Role of the Government ?

The most powerful organization in any country which can turn prevalent trends towards empowerment is the government. In this regard it is our hope that the government introduce or streamline the following areas :

Provide Technological Support

Technology is for every strata of the society and women have to be an equal beneficiary of the products and processes which emerge from it use. The range of areas in which technology can put a greater control in the hands of women is extensive, and constantly growing, from managing distribution of natural resources at the village-level to having access to political leadership opportunities. Technology in conjunction with development opportunities has the potential to reach those women who up till now have not been contacted by any other means, thereby empowering them to participate in economic and social progress, and make informed decision on issues that affect them. During the course of our research we were able to identify three areas of technology research and adaptation, which could make substantial contribution to rural women's welfare and empowerment, viz., agricultural production and post-harvest processing, information technology, and energy. We have endeavoured to provide examples from all over the world to illustrate these empowerment areas.

Agricultural and Post-Harvest Technologies

During the 1980s and 1990s, agricultural research of the Consultative Group on International Agricultural Research (CGIAR) brought new visibility to women's roles on the farm. For sixteen years, researchers allied with the Women in Rice Farming Systems network based at the International Rice Research Institute (IRRI) carried out sustained research with women farmers throughout South-East Asia. They developed both the technologies that addressed women's specific interests and knowledge and the means for conducting gender-sensitive participatory research. Technologies—improved germplasm, crop management and cropping systems, and machinery—were tested and evaluated in specific rice-farming environments using gender analysis, participatory methods, and a systems perspective by a multidisciplinary team of researchers.

In West Africa, scientists at the West Africa Rice Development Association (WARDA) used biotechnology to develop a rice variety with the high yields and shatter-resistance of Asian rice and the drought resistance and broad leaves of African rice. The new variety substantially reduces weeding requirements, lightening a task usually performed by women and children. Despite such initiatives, national-level progress has been slow and more attention to women's needs is required. As more men migrate to urban areas, family labour becomes scarce, soil fertility declines, and more poor rural women become farm managers, women need labour-saving and knowledge-based technologies to improve the productivity of both land and labour.

Source: Paris, Thelma, R., Feldstein, Hilary, Sims and Duron, Guadalupe (2001) Technology, http://www.ifpri.org/2020/focus/focus06/focus06_05.htm

Information Technology

Information Village set up by Swaminathan Foundation

Village knowledge centres (earlier known as 'information shops') have been set up in three places in Villianur district in Pondicherry. A gender expert was invited to participate in the inception stages to ensure that gender sensitivity was built into all the operations. All the identified volunteers and the project staff were also given orientation to the importance of incorporating gender sensitivity through a workshop.

Training of Volunteers

Training	*Duration*
Gaining familiarity with basic operations of windows 95	2 Weeks
Transaction of data on wireless	3 Sittings
Gaining preliminary knowledge of HTML	1 Week
Learning of Word 97	2 Days
Learning of Power Point	1 Week
Use of Windows 95 keyboard for Tamil Fonts	10 Days

The agricultural labourers, especially women whose wages are partly derived from grains thus get to know the sale prices of the grains through the net. Here they also get health related information. The I-centres also provide them information related to training in new skills to augment their income, and on public schemes on rural welfare and the government's list of eligible families living below the poverty line. An analysis of users' registers maintained in the village knowledge centres reveals that female users form 17.24 per cent or 1388 out of 8047 of the total users

Source: Balaji, V.; Rajamohan, K.R.; Pandy, Rajasekara; Senthilkumaran, S. (1999). The Information Village Experiment in Pondicherry, itt.nissat.tripod.com/itt20004/infovill.htm

Energy

Biogas Plants

They are the answer to fuel and power related problems in villages. SCOPE succeeded, where government failed by, creating among beneficiaries awareness on the importance and need of such units. The saga continues even today with differences. A total of 15 biogas plants have been constructed in the target area of which 4 are connected to latrines. This saves village women precious time for collecting fuelwood, sizable quantities of cow dung. It is expected that more families will construct biogas plants. As a result of latrines being connected to bio-gas units the problem of leach pits filling up has not been witnessed even after a long time, The gas produced can be used for lighting and cooking resulting in conservation of energy, reduction in expenses and continuous availability of energy for lighting and cooking. Further the slurry can be used as a fertilizer, resulting in reduction in use of chemical fertilizer and additional income by selling it (or) reduction of expense on purchase of manure.

Source: SCOPE TRICHY, 2003, http://www.scopetrichy.org/biogasplants.html

Guarantee Participation

Community development practice is primarily concerned with building social and economic development institutions that lead to community self-sufficiency. Not surprisingly, women in rural India today finds herself at an important political, social, and economic juncture. Under present development environment, government agencies, designed to promote economic and social independence of the rural poor, must assume a more entrepreneurial posture if the growth rate is to survive the challenges that confront it. However, the economic and social goals of development are increasingly in conflict as government resources shrink under various cost cutting and rigorous accounting pressures. It has, therefore, become an uphill task to balance these two goals, since both social and economic goals suffer from the lack of resources and the lack of skilled practitioners to organize and run the community development initiatives. In this regard, the governmental agencies can modify their role to provide customized information products for specific inquiries with the help of vast Field Publicity networks and numerous Panchayats spread across India. Trained rural workers can guide the locals populace in issues relating to economic recovery, local government planning projects, rural health issues, funding sources, and other related issues for the purpose of monitoring the quality of rural life. An illustration to the same has been provided to further strengthen the line of reasoning.

Women's Empowerment through 'Thrift' and 'Credit'

Siliguri, Assam: For the past two decades the government has taken up numerous programmes for the economic uplift of women below the poverty line and also for their empowerment. But as it happens with most such schemes, these well-intentioned steps do not bring the desired results for the deprived sections. An exception is the Swarna Jayanti Saahari Rojgar Yojana (SJSRY) one of its components, "Thrift and Credit" encourages the formation of women's thrift and credit groups with 10-20 members bringing their daily savings to banks where accounts are opened in the group's name. The savings thus made normally does not exceed Rs. 20 per month per family, sometimes even less. But it is surprising how these little drops have made a mighty stream. These women join the SJSRY through the process of a survey weeding out families earning more than Rs. 313.12 per head per month. Women from families of small vendors, rickshaw or van pullers, fruit, fish, vegetable vendors, newspapermen, milk vendors, domestic helps and women members of families of other professions doing small jobs are the main constituents. They number about 33,000 families and one woman from each family forms the SJSRY brigade in the Siliguri Municipal Corporation. About 1,216 Opinion Leaders, designated as Resident Community Volunteers (RCVs), from these members, have been trained and motivated by the Ramakrishna Mission, Lokasiksha Parishad, Narendrapur. The mission has put before them a dream. They were motivated to form thrift and credit groups, pool their smallest savings and deposit them in banks. This inculcates the thrift habit, puts savings of poor people into mainstream banking, makes provisions for urgent needs, augments the group's own capital base, enables members to start small business units and, above all, instils in women a confidence in themselves. Almost 500 Thrift and Credit groups have come up within the Siliguri Municipal Corporation with an average of 20 members per group covering 9,000 families. The total savings from all these groups have touched an unbelievable Rs. 62 lakh and with government contribution of Rs. 53 lakh towards a revolving fund bringing the total group capital to about Rs. 115 lakh within five years. Newly 4,000 members got loans from their own capital and are doing business at nominal interest. The most revealing aspect is that the percentage of loan recovery is an amazing 120 per cent. All the work connected with running of these groups is being done by the group members themselves from collecting monthly savings from members, depositing them in banks, withdrawing the amount when necessary, deciding the viability of the business their members want to start, sanctioning of loans and ensuring timely repayment in instalments. Through these group activities they are experiencing a sense of belonging with the mainstream society and are acquainted with the outside world. The municipal corporation authorities, realizing the potential of such a workforce, have started decentralizing some of their earmarked works. These include identifying beneficiaries for Antyodaya and Annapurna Yojana, Balika Smriddhee Yojana, Old Age Pension, sponsoring of loan applications for micro-credit from banks for the BPL population, identification and prioritization of small schemes like road repairs, drainage repair, installation of roadside water stand for slum dwellers and installation of community latrines and bathrooms. The corporation has also innovated schemes for development of the slum children. It had organized annual sports. About 3,000 have participated in different sports with prizes galore, opening an avenue for the deprived and neglected. It has started league football with 15 teams. A visit to these CDSs and their members will make one feel the level of interest, enthusiasm and confidence among the women who are running the show. The scope before the CDSs for undertaking other development work is immense. Members are now ready to take on various schemes of the government. Proper implementation of these with the help of this large workforce can be of invaluable help, this in turn will help the people become part of the mainstream. The SJSRY programme is now run in most municipal corporations where the elected leaders—the councillors—are motivated.

Courtesy: *The Statesman* in Grameen Bharat, Vol. 3, Issue No. 8.

Ensure Executive Reformation and Enhanced Group Effort

Sustainable development in India has to be routed through e-Government strategy and open source policy. In other words, the executives need to enable construction of public infrastructure such as the national and local governments WAN and a comprehensive development of electronic filing system, amendments in the requisite laws and the constitutional safeguards to handle electronic data in public and private procedures in the same way as paper documents, government procedures be made available on the net. Such steps in turn will pave the way for one stop electronic filing system for government services, the business process restructuring of internal procedures of the government, an innovative work-flow reorganization and a comprehensive IT systems' introduction, designing, procurement, and operation methods.

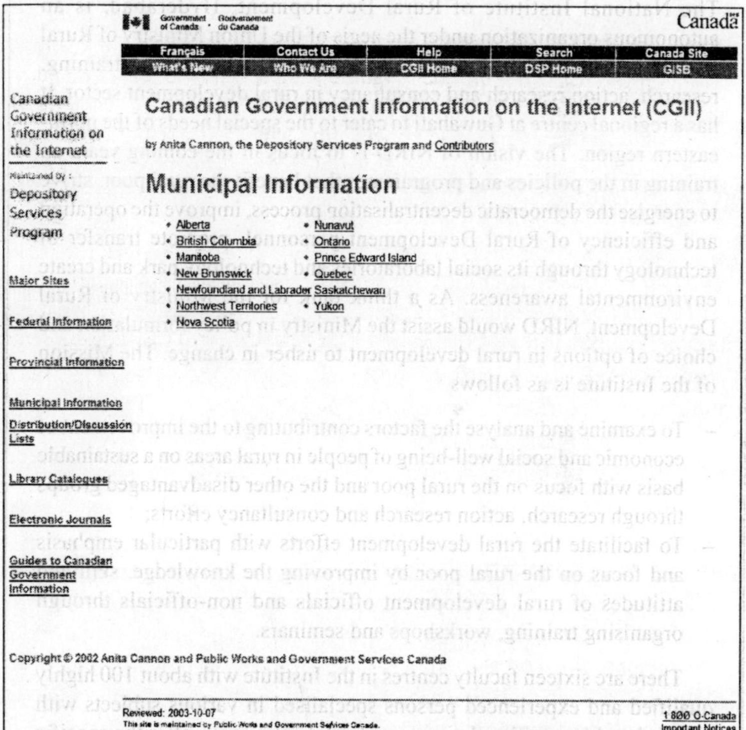

Fig 9.1 : Candian Government and Municipal Informations on the Internet

Source: Government of Canada, http://cgii.gc.ca/muni-e.html

This emerging scenario requires adept workers, however, administrators report that it is difficult to change attitudes and expectations of the existing staff as developmental processes have moved from distribution to active involvement. Sifting, selecting those that are good at paying attention to details, managing deadlines and completing technical tasks, and training with the help of sensitizing scenarios, multidiscipline and indepth professional development tutoring, at this juncture thus makes a lot of sense. Given below is an example of one such facility. The affectivity of this institute can be further enhanced by increasing its reach to the village units of the country.

Training

National Institute of Rural Development

The National Institute of Rural Development, Hyderabad, is an autonomous organization under the aegis of the Union Ministry of Rural Development. It is the country's apex body for undertaking training, research, action research and consultancy in rural development sector. It has a regional centre at Guwahati to cater to the special needs of the north-eastern region. The vision of NIRD is to focus in the coming years on training in the policies and programmes that benefit the rural poor, strive to energise the democratic decentralisation process, improve the operation and efficiency of Rural Development personnel, promote transfer of technology through its social laboratories and technology park and create environmental awareness. As a think tank for the Ministry of Rural Development, NIRD would assist the Ministry in policy formulation and choice of options in rural development to usher in change. The Mission of the Institute is as follows :

- To examine and analyse the factors contributing to the improvement of economic and social well-being of people in rural areas on a sustainable basis with focus on the rural poor and the other disadvantaged groups through research, action research and consultancy efforts;
- To facilitate the rural development efforts with particular emphasis and focus on the rural poor by improving the knowledge, skill and attitudes of rural development officials and non-officials through organising training, workshops and seminars.

There are sixteen faculty centres in the Institute with about 100 highly qualified and experienced persons specialised in various subjects with national and international exposure. Each centre looks after the specific programme of the Ministry on which a group works in the respective programme areas. The Institute has well-equipped infrastructural facilities to conduct the programmes.

Source: Annual Report 2002-2003, Ministry of Rural Development, Government of India.

Respond Professionally to Growth Challenges

Long-established community restrictions have curtailed women's involvement in the economy and their access to resources. This is now compounded with international concerns of balance of payments, structural modification agendas, diminishing stipulations of commerce, and war. Severe measures by the government has reduced public spending, adversely impacting the social sector programmes and causing hardship for rural women and their dependents. Against a dismal backdrop the focus is now shifting towards assisting women to secure income through their own efforts. Women's access to credit at affordable rates was found to be a major hurdle in this area. This situation throws up a challenge as to whether the Government should shift its focus away from 'income-generating' activities or concentrate on paradigms of, and support for, women's productive activities, which may operate in tandem with income-generating activities. Caught in this dilemma, the government is taking a long time to focus on an obvious answer, viz., (1) Participatory Approaches, and (2) Approaches that Introduce potential beneficiaries to various credit opportunities.

Participatory Approach

The Self-Employed Women's Association (SEWA), Gujarat (India)

SEWA is an autonomous trade union for women which began in 1972. It has both urban and rural members and originated from the women's group within the Textile Labour Association. It has, since its inception, gone beyond conventional trade unionism. This has been done in response to the inability of conventional unions to deal with the problems of poor women in the informal sector. It is based on the desire to establish recognition for self-employed women. From this basis it has built components responding to its membership's demands. These have included credit, skills training and a wide range of responses to social problems such as maternal protection schemes, widowhood benefits, child care and training of midwives. In 1974, SEWA registered its own bank. It began with members purchasing small shares. The process of building the bank required "close supervision, self-monitoring through participatory process, and building a collective consciousness and cooperative values among the SEWA groups. Training and technical inputs to improve the efficiency of activities were also necessary" (Wignaraja, 1990: 75). The bank has adapted its operations to the needs of illiterate women. Loan applications are filled out by bank workers based on face-to-face discussions. Data provided by the applicant are confirmed by a field visit. The time between initial loan applications and disbursement is short. Loans are repaid in 36 monthly instalments. The repayment rate on loans from the SEWA bank is 87 per cent.

Source: Albee, Alana (1994) *Support to Women's Productive and Income-Generating Activities*, http://www.gdrc.org/icm/wind/wind-unicef-wp.html#KEY

Approaches that Introduce Potential Beneficiaries to Various Credit Opportunities

The Agricultural Development Bank : Nepal

The Agricultural Bank of Nepal provides an example of an existing banking institution that is extending facilities and modifying procedures in order to serve the poor. Its Small Farmer Development Project (SFDP), supported by loans from IFAD, is the administrative umbrella through which small loans are channelled to farmers and the landless in a number of development initiatives financed by the Government and different international agencies. The SFDP administers the funds to various projects, and the bank's branch offices deliver loans and collect repayments through their fieldworkers, the group organizers and women's group organizers. The group organizer is responsible for visiting villages, conducting household surveys to identify those interested and eligible, and for forming groups. By 1988 there were 532 women's groups providing loans to over 4,000 women, compared to 4,667 farmers groups with 42,345 members, mainly men. Eligibility criteria used in screening loan applicants has been designed to favour the rural poor, both men and women. The small size of the loans provided for groups or individuals are within the women's borrowing capacity, and the activities approved include those that are attractive to them. Collateral requirements have been replaced by group liability and peer pressure for repayment. There is a gradual shift from collateral-based lending to a development orientation and project viability in order to cater to the needs of the weaker segments of the population, who have a low resource base.

Source: Fong, M., and H. Perrett. (1991). *Women and Credit*, FINAFRICA.

Initiate, Support, Develop, and Collate Gender Sensitive Statistics, such as school drop-out rates, single households, credit receipt capacity and length of unemployment, which can eventually lead to gender responsive budget initiatives

Gender sensitive statistics are important if one is to address the critical issues of feminization of poverty as it serves the function of sensitizing the policy makers to design effective strategic solutions, that enhance the economic options of women.

Ensure Gender Scrutiny of the Budget

Analysis of such nature help examine the allocation of resources among the population. They assist in developing mechanisms with which women's contribution through unpaid labour can be measured. Such an in-depth scrutiny often helps in gauging gender priorities and in designing suitable techniques to track improvements with the help of collated data.

Formulate and Operationalize an Equal Opportunities Plan

This aspect particularly relates to the need for legislative, administrative and fiscal measures to ensure equitable land rights for women, support for home based workers by strengthening their social protection cover, and access to competent health facilities. Sporadic activities like the one tabled below are not an answer to an endemic problem that is reflected in the laws laid down in the Constitution of India.

Centre Scans Gender Bias in Laws

New Delhi, December 19—Prime Minister Manmohan Singh has asked the Law Minister H.R. Bharadwaj to ensure gender-equitable legislative regime. In keeping with this directive, the Law Ministry has reportedly started the process of scanning laws for gender bias in them.

Various women's groups and organizations, including the National Commission of Women, are being contacted for inputs and suggestions to help streamline legislations which are not gender-sensitive.

The exercise ensuring gender equality in legislations is in keeping with the commitment made by UPA in Common Minimum Programme. The document, among other things, assures complete legal equality for women, especially by removing discriminatory legislation and enacting new legislation. The Cabinet's clearance to amend the Hindu Succession Act so that women also get a share in their father's property is part of the move to make laws equitable. So is its decision to amend the Prevention of Child Marriages Act.

Child marriage is a cognizable offence and parents and priests solemnizing these marriages are also liable for punishment according to the proposed changes in the law. Similarly, children involved in such marriages may have the choice of opting out of it. In areas where child marriages are rampant, 'councillors' would be appointed to create public awareness against the practice.

Women groups too have highlighted the need to review existing legislations. The NCW had, in recent years, forwarded to Centre a list of 42 gender related laws, which needed correction.

The need to have a re-look at some of these laws has often been underlined by NCW's Poornima Advani. In a meeting with the Chief Justices of High Courts, she pointed out that the Indian Penal Code has been amended only twice since it was enacted in 1860. "It is a Victorian law," she said. She also sought a redefinition of terms like bigamy and rape.

Given the opposition of allies like the RJD, the Centre may find itself in a loop when it comes to bringing the Women's Reservation Bill in Parliament. The Domestic Violence (Prevention) Bill and the Bill against sexual harassment in the work place too have been hanging fire for quite some time.

Saroj Negi, *Hindustan Times*, 20th December, 2004, pp. 13, New Delhi.

Buzzword: Equality

- UPA's common minimum programme had promised legal equality for woman.
- Law Ministry will check all laws for gender bias.
- To seek inputs from NCW, other women groups.

Child Marriages	**NCW Wants** Changes in 42 laws including Cr.P.C., IPC, Indian Evidence Act, Dowry Prohibition Act, Commission of Sati (Prevention) Act, Hindu Marriage Act, Indecent Representation of Women (Prohibition) Act, Immoral Traffic (Prevention) Act, Foreign Marriage Act, Representation of People Act.
Punishment: Parents, Priests solemnizing child marriages to be penalized. **Remedy:** 'Child Couple' may have option to opt out of marriage later. **Prevention:** Councillors to be appointed to create public awareness.	

Source : *Hindustan Times*, 20th December, 2004, p. 13, New Delhi.

What should be the Role of the Corporates and the Non-Governmental Organizations?

These institutions are a repository of intense human will/ manpower, social reach, technology and financial clout. At the same time, India is perhaps the only country in the world where the social and the corporate goals can exist in harmony. Various indicators like world's largest arable land mass, immense consumption base, diverse climatic zones, steadily increasing rural income, tax incentives and a proactive policy vision, etc., point towards the efficacy of this fact. It is against this background that we hope that they enlarge and fine tune their operations in the areas of sustainable women empowerment by augmenting the potential and leadership of women's organizations.

Unknown to most of the women, a woman is a born leader for she exercises considerable influence in her immediate environment. It, therefore, falls on various facilitating agencies to assist her in recognizing that how as a leader can she relate to those she leads. What does she thinks about herself and about others, and how does this influence, how she treats others and help her to understand that leadership is about doing the right thing and not doing the wrong thing well. It is about taking responsibility. Just as new recruits in an organization develop with time, experience and mentoring, rural leadership is also about transformation and a change in attitude and behaviour, which takes time. It is in the corporate firms best interest to stick with the process of identifying talent and encouraging women to be the best, because the world is not made up of just men; women must take up the challenge and play their role. This is all about changing the mindset of women so they do not view their successes through the eyes of men. The challenge for corporates and NGOs is to stimulate women to identify with women candidates.

CCF Promotes Women's Enterprise Development among Rural Communities in India

During EVP Peter Woicke's recent visit to India, IFC and Usha Martin Limited (UML), India's leading manufacturer of steel ropes and wires, signed an agreement to promote women's enterprise development among rural communities in India's Jharkhand State. UML is the principal corporate sponsor of Krishi Gram Vikas Kendra (KGVK), a non-governmental organization established in 1977 to improve the livelihoods of Jharkhand's rural poor. KGVK implements health, education, agricultural extension, women's empowerment, and natural resource management initiatives in 115 villages in and around the cities of Jamshedpur and Ranchi. Many of Jharkhand's rural poor belong to India's minority tribal population. The 15-month programme – jointly funded with contributions from UML and IFC's **Corporate Citizenship Facility** – will link KGVK with SEWA (Self-Employed Women's Association), India's most successful model of informal sector enterprise development for women. SEWA will provide training and technical support to KGVK to strengthen its capacity for identifying and promoting enterprise opportunities for rural women. The programme will continue to support the implementation of a number of pilot projects that, if successful, will be integrated into KGVK's ongoing development work. SEWA is a membership group that helps women achieve self-reliance through full employment at the household level. From its base in Ahmedabad, SEWA has organized over 500,000 women representing 74 different trades and occupations in the informal sector. Having recently become familiar with SEWA's success, UML's Chairman, B.K. Jhawar approached IFC for assistance in "replicating" SEWA in Jharkhand to build on KGVK's existing initiatives in women's enterprise development with SEWA's expertise in organization, micro-credit and marketing. In so doing, UML seeks to expand the employment opportunities and, thereby, improve the livelihoods of rural women who are KGVK's constituents. IFC's SME department – through its Capacity Building Facility – is currently providing technical support to SEWA to expand the production and sales of handicraft embroidered goods made by its members through the creation of a trade facilitation centre. This initiative represents a novel approach to linking poor rural artisans to international markets through a vehicle that is owned by the artisans themselves. SEWA's guidance of KGVK efforts to promote women's employment and enterprise in Jharkhand may foster similar linkages between the SME Department and KGVK to test such initiatives in Jharkhand. For example, UML currently contracts with a dozen cottage industries operated by women with the support of KGVK for the manufacture of small items used in the manufacture of steel rope and cable in Jamshedpur. These items include works gloves, cotton gaskets and identification tags for inventory control. CCF supports UML's efforts to increase employment and the diversification of local economic opportunities for rural women. These efforts, and their anticipated outcome, will point the way for Indian companies who, like UML, seek to move beyond philanthropy and provide sustainable economic opportunities to its neighbouring communities

Source: http://www.indiapartnershipforum.org/pdfs/ifc_india.pdf

Affect Opinionated and Fiscal Support for Women from a wide range of Stakeholders

Often women enter the market as unequal participants and often leave at the end of the day with unequal rewards. So for the markets, for women have been synonymous with poverty, inequality and large scale unemployment. NGOs and the Corporate have their tasks cut out for them in that they are urgently required to turn the public opinion in favour of women and their produce. Adverse and redundant public perception towards women's role and empowerment has taken its toll in the form of socio-financial insecurity among women. The solution though simple would take some dedicated effort from the Corporates and NGOs and that is to create women friendly markets. Such markets would allow the rural women to participate holistically in their (market's) operation and have an equitable share in their profits. The following areas need to be addressed if this issue of women friendly markets are to be resolved in the favour of rural women :

1. NGOs and the Corporates need to prepare women by
   — *Providing and facilitating, education and skills*: This will create awareness about their legal rights, market operations and forces that impact upon it, it would also help them to develop their priorities and incentive systems, besides providing them with freedom to act in accordance to their benefits while maintaining competency and quality awareness.
   — *Ensuring access to credit, information*: This would assure financial flexibility, budgeting, cost competency and confidence to handle demand shifts in the product category in which they deal. It would also help them to be on guard against any financial malpractice that they may encounter
   — *Physical infrastructure*: This will provide visibility, permanency, effective time management, and speedy turnover, besides ensuring access networking /business opportunities, regular buyers, profitable markets and other social safety nets.

Corporates and NGOs can use their clouts with the government to ensure market and environment protection at the initial stages for these women, by pushing for legislations, (A) that ensure clean air, water, soil and ample forest areas; (B) that ensure financial, social, legal and physical protection from numerous unlawful activities that often target employed and self-employed women; (C) that provide dedicated markets for the products and the produce for the women; and (D) that ensure the networking of diverse groupings and interests, so that tension arising due to change in gender relationships and the assertion of traditional rural cultural values can be effectively resolved.

Fashion New Alliances/Networks with Self, other NGOs, Educational Institutions and Government

Increasing information levels, awareness and exposure to market forces is instrumental in an expression of the desire of women in agriculture/villages for recognition and visibility and for a greater role in decision-making. A networking movement that encompasses the diversity of rural women's lives and their environment, and addresses a wide range of social as well as economic issues, is a goal which can be achieved if the corporates use their immense clout with the government agencies and the market. Such an involvement in creating rural women's networks can assist in training in leadership resulting in growth in confidence which will assist the women to participate in decision making in the public arena, besides increasing good will, areas of influence and lucrative markets for the corporates. In terms of the benefits to women, such networks can provide many advantages for individual participants, like building up a capacity among the women shareholders to promote the broader social movement for change in the status of women.

Industry Partnerships Investing in Rural Women

The Industry Partnership – Corporate Governance for Rural Women Programme is a joint government and industry initiative. It aims to provide the expertise for rural women to develop their skills and increase participation in industry. The programme will also help industry organisations tap into a wider range of experience and expertise, thereby strengthening our rural sector. Its role is to improve recognition of the contribution women, young people and indigenous communities make to rural industries (agriculture, fisheries, resource management, food processing and forestry) and to encourage them to become part of the decision making process in their industries. To achieve these goals, it designs and implements programmes that provide opportunities for people in these client groups to develop skills and network. The new skills and networks thus enable them to contribute effectively to their respective industries and communities. It also uses mentoring as a technique to enable an individual to share professional and personal skills and experiences, and contributes to career growth and development. Through mentoring relationships, the mentor becomes the mentoree's source of advice, guidance and knowledge for professional and personal development.

Source: http://www.rirdc.gov.au

Given below are three exemplars which can be utilized by NGOs and the corporates to accomplish this task—

Exemplar 1. The networking among diverse women organizations and organizations like the educational institutions, other NGOs, and varied industry and informal groups can be coordinated by salaried staff located in a special unit within a designated area in the district headquarters, governed by a consultative committee drawn from among various women stakeholders and the representatives of the networked organizations. Operational costs including salaries can be funded *via* programmes, *haats* and other fund raising activities that organized by the governing committee.

Exemplar 2. This model is community-based, usually governed by an incorporated community organization and maintained by volunteers. Such networks can be funded by member contributions which can be supported by financial or in-kind support from government and industry sponsorship.

Exemplar 3. This is an industry-centric alliance of farm women, wherein (a) the establishment of a network of women's groups is targeted to enhance an industry-specific producer culture and thus the profitability of the industry or (b) the women's group which though initiated by women in the rural community, is closely associated with and supported by a corporate organization.

Accept pilot assignments to try ground-breaking developments in women's empowerment and gender integration in the society. Constructing a knowledge base on operative policies for stimulating middle-of-the-road projects

This acts as a pointer towards a growing need to unleash technology programmes focused at women at grassroots level. The concept of targeting/involving women is based on the belief that training women implies training the whole community. At the same time corporates in conjunction with NGOs should explore various business techniques to get to the root cause of critical issues within a community of which women form an important part. The data collected in such a manner unveils nascent requirements for women empowerment. It also provides the foundation of knowledge that can be leveraged to assist the constituency of women around the area and with it around the world to enjoy the benefits of technology.

The Global Citizenship Initiatives of Hewlett-Packard

US Digital Villages were our first major e-inclusion projects and were designed to help underserved communities in the US to use technology for learning, working and playing, and to make links within the community and beyond. In each case, the communities worked with HP to develop and implement a Community Technology Partnership Plan with a shared vision and strategy to meet those needs, including a plan for sustainability of the programme at the end of our three-year funding. In the US there are three projects: the Baltimore, Maryland Digital Village, the East Palo Alto, California Digital Village, and the Southern California Tribal Digital Village. To date, more than 5,000 residents in Baltimore have used community technology centres to learn computer skills and receive job training. The East Palo Alto Digital Village delivered technology to more than 70 small businesses and entrepreneurship training to another 100 businesses. This has generated $ 2.75 million in revenue for these businesses and 157 new jobs since the project's launch in 2001. **i-communities** are development initiatives located in emerging markets and executed in partnership with local government, non-governmental and community organizations. i-communities use technology to promote sustainable social and economic development. Specific solutions focus on increasing literacy, raising income levels through entrepreneurship and job creation, and providing access to government, healthcare and education services. As targeted communities benefit and grow, HP learns how to compete in these markets. Based upon this knowledge, HP introduced a multi-user desktop solution that enables educators to maximize limited budgets while integrating technology into the classroom. We currently have i-communities in Kuppam, (Andhra Pradesh State) India, Mogalakwena, (Limpopo Province) South Africa and Houston, Texas, US. The Kuppam i-community celebrated its first anniversary in February 2003 with the launch of a mobile information unit that reaches more than 12,000 people in 150 outlying villages each month. The project has provided ICT infrastructure that gives communities expanded access to government-to-citizen services as well as agricultural information and services such as soil testing to improve crop yield. **HP's pilot solar-powered mobile photography studio has enabled women from local self-help groups to become entrepreneurs and double their family income. These solutions are designed to become economically self-sustaining.**

Source: http://www.hp.com/hpinfo/globalcitizenship/gcreport/socialinvest/einclusion.html

What should be the Role of Community?

Gender development is influenced by community norms and values. This is to say that community is a strong predictor of women's empowerment. The relationship of both community and an individual woman's traits to different measures of empowerment vary, suggesting that "empowerment" is inherently a multi-dimensional phenomenon, with women relatively empowered in some spheres but not in others (Mason and Smith, 2003). Key thrust areas wherein communities can contribute effectively have thus been identified as:

Communities can play an important role in enhancing the power of women within the household, because these powers are determined by social institutions and not by individuals. Women empowerment is multidimensional and socially interrelated, therefore, it is largely dependent on social contexts in which it is applied

Acknowledgement of the women's efforts in the household operations specifically in the area money management is of utmost importance because this quality is often instrumental in her being able to provide all the basic amenities in her home. With limited earnings she has successfully looked after the needs of education for her children, health needs of her family and the ceremonial needs of the society. The quality of care and concern when focused on the societal needs evolves into a a controlled power position which needs to be harnessed. This can only be done if the communities accept her contribution and learns to respect it. The illustration given below exemplifies this aspect.

From Maid to Sarpanch

Bhilai, December 27, From scrubbing dishes for a living to running the affairs of an entire village, it is a giant leap for Rahi Mehar. The domestic help is all set to become sarpanch of Kodia village panchayat next month,

> without even contesting an election. Villagers are unanimous in their choice of who will lead their panchayat. The post of sarpanch in the village had recently been reserved for Scheduled Castes, and Mehar's is the only SC family in the village of late. Mehar has been spending a lot of time in understanding the problems of her village. "Initially, I thought the people were just making fun. I have been working in houses in Bhilai for the past 16 years. So, it was not easy to believe. Later, I came to know that none was eligible for the post," Mehar said. Privation had forced her to work as a child labourer. Later, she worked as cleaner for two years before some one introduced her to few families in Bhilai, who were looking for domestic helps. Mehar says she has yet to learn the functioning of the panchayat and its duties, but adds, "I will be more successful than those with years of experience". "I don't know much about development work but I know the deteriorating politics everywhere. My villagers know that I'll not play politics at any cost. They told me that roads are most important and I can tell those watching me to come again to my village after six months and see the quality of the roads here", she adds. **Piyush Srivastava**
>
> *Source*: *The Hindustan Times,* December 28, 2004, p. 9, New Delhi.

The society/community needs to change its definition of bench-marking various development milestones of the women because these milestones index only selective aspects of women's empowerment and fail to provide a complete picture

Often communities target outreach initiatives by women and by outside agencies which propose to bring basic access to technology and community development programmes to communities, citing reasons of way of life, cultural constraints and rituals. Often lack of awareness is not the key issues here. Inherent fear in the society regarding the modification of the social norms is what lies at the base of this act. It manifests itself in the factors contributing to social and economic exclusion included rural isolation, domestic isolation, dependency and family pressures, stereotypical attitudes to women's roles, social policies that do not recognize women's needs and decision making processes that deny women a voice. It is, therefore, important that approaches that are aimed at capacity-building projects are designed to catalyze economic and social development of rural women through investments in rural organizational infrastructure be expressed/sold/ approached by the catalyst groups in a more ambitious manner so that the communities become more amenable towards them. Once they realize the efficacy of these measures, it will not be difficult to convince them about the Entrepreneurship and education potential of these projects which will ultimately culminate in leadership development and individual skills building for women. Critical awareness can thus be developed through education provided by facilitators at the grassroot level. Such focused approaches can be leveraged to change the milestone index of women empowerment which is still linked to marriage, kids and duties towards the household and the extended family.

The Role of Women in Rural Electrification in the Philippines

Developing a community renewable energy system takes the dedication of an entire village. Everyone, men and women, must work together to bring the project into reality. Recently the Yamog Renewable Energy Development Group, Inc. in partnership with the people of Polocon, an off-grid farming village in the South-East Philippines, completed a 15 kW micro-hydro system that generates electricity for 525 people in 105 households. The women actively participated in this important community development initiative and their involvement proved to be invaluable to the success of the project. Representatives from an existing women's organization in Polocon serve on the village electrification committee. This allowed women the opportunity to directly influence the planning, management, construction and management of the microhydro project. Nay Elie, a woman leader in the village said, "The entire community acknowledged the positive presence of women on the committee and their effective leadership abilities". From conceptualization through the evaluation phase, the interests and concerns of women were taken into account. The needs of women—better lighting, being able to do other household tasks the evening hours, improved educational opportunities for their children—were all incorporated into the project design. Now, an electric corn mill and coffee hull lighten the women's workload and all of the women appreciate the fact that with electricity their lives are more comfortable. Women participated in the actual construction of the system itself, hauling sand and gravel to ensure the project's completion. They helped replant the watershed with indigenous trees to prevent erosion and to ensure an adequate supply of water for long-term sustainability of the system. Women members of the village electrification committee also collect the payments for the electricity at a rate set by the committee and do the book keeping. One day, a woman leader in Polocon, summarized the results of women's participation in the project best by commenting: "The women of Polocon are highly regarded among men and women alike, as they contributed greatly to the realization of this project towards the increasing sustainability of our community".

Source: November 2001 Green Empowerment serving people and saving the environment with renewable energy. http://www.greenempowerment.org/media/greenemp.pdf

Health Care Management : Some Conclusions

Health and access to treatment cannot be considered in isolation from the management of health systems in the country, national development and health policy. Besides, access to health services is not only driven by the availability and quality of health care services, health personnel and medicines but also by the costs of treatment and medicines, and by the availability of functional, locally adapted health institutions along with the services, competence of health personnel, the delivery system and the cogent use of quality medicines, and the economics of the health services. At the same time, we cannot do away with socio-cultural behavioural patterns and educational background of the beneficiary. Lack of financial resources among the rural population acts as a chief barrier towards the national goal of achieving a disease free society. Since many people in the country are unable to pay either for diagnosis and treatment at health centres or for drugs, this issue has evolved into a major area of concern. According to Novartis Foundation, in the mid-1980s, as a consequence of structural adjustment policies, the introduction of cost-covering policies in the health sector in many developing countries, including India, exacerbated the issue of access for the poorest sections of the population even more.

The United Nations Development Programme (UNDP) estimates, some 20 per cent of people in developing countries are excluded from access to health services, i.e., health services are either not adequately available, as in many rural areas or treatment and drugs are not accessible due to lack of financial resources. As high as 51 per cent of the rural population belonging to the least developed countries is excluded from the ambit of health services. World Health Organization's (WHO) Commission on Macroeconomics and Health has emphasized the importance of health and access to health care, both for economic and social as well as individual development. The authors consider promoting health in the developing nations as a primary means of combating poverty. Needless to say that "the disease burden, expressed in terms of morbidity (cases of disease) and mortality, is slowing down economic development" (Novartis Foundation, 2004). Better health for people in developing countries is a prerequisite for economic and social development and it should not be considered a byproduct of economic achievements of a nation.

Individuals living in poverty, equate disease with low work less productivity, especially when they earn their living from occupations like working in the fields, collecting wood, manual work, etc. It is particularly devastating, since it is almost impossible for them to compensate for disease-related losses of income. It must be understood that poverty affects nutrition, sanitation and housing, which in turn makes people more susceptible to infectious diseases. It is not surprising that poor people avoid necessary treatment or postpone it as it stretches their already meagre finances. Poor are thus caught in a vicious cycle of underdevelopment wherein poverty causes disease and disease exacerbates it.

The authors are of the opinion that following areas need particular attention from health authorities (both public and private) and the policy makers.

Infant, Child, and Maternal Health

Due to highly deficient nutrition, unhygienic surroundings and pollution infants, children and women in developing countries, suffer from infectious and poverty related diseases, viz., diarrhoea and parasitic diseases like diarrhoea, gastroenteritis, Tuberculosis, etc. Consequently, there exist high episodes of infant, child and maternal mortality. This in turn has a particularly severe and negative impact on the affected family. Because (a) the focus then shifts from the birth to the survival of the child, illustrated by high birth rates (increase in the population) in the country. This factor though comprehensible at people-to-people level spells disaster at the developmental level, and (b) Short intervals between two births implies that a woman gives birth more frequently. which is to say that, year after year the health of women continue to deteriorate at a rapid pace. Authors advice that the current health policies be augmented by highly trained health workforce comprising of doctors, hospitals, nursing and village level health workers.

Structural Deficiencies in Health Services

Health problems in the developing country like India can be attributed to, besides other factors, deficient and overloaded health services. The authors have tried to identify these problematic areas which need a careful 'looking into' by the health and policy authorities:

- Lopsided or Improper Resource Allocation.
- Inequitable access to health-care services.
- Ineffective utilization of funds.
- Cost Escalation.

Lopsided or Improper Resource Allocation

Developing countries often have issues regarding proper funding of numerous public welfare programmes. Balancing them with the other needs of the country, viz., defence, external affairs, administrative affairs often requires skill of a magician. Allocation of funds to various public programmes depends upon the political inclinations of the ruling political parties. Expenditures such as that of defence often eat away into the limited resources of a developing nation.

Inequitable Access to Health-Care Services

Politicians and senior government officials often like to surround themselves with expensive specialist medical facilities which again are situated only in the urban areas, and are frequently inaccessible to the masses, and meagre resources often go into establishing them, while cost-effective interventions such as comprehensive basic health services and provision of essential medicines lapse due to lack of resources. Needless to say that this mainly affects the 'have nots' of the society, whose need for vital health are often more pressing/critical.

Ineffective Utilization of Funds

An important observation in this regard would be about the utilization of resources available for the purchase of medicines for public health services, where inadequately informed government personnel end up purchasing proprietary medicines instead of generics.

Cost Escalation

Bureaucratic red tape, bungling, corruption and the lack of human will and misinformed population all contribute to the increase in the cost of various rural health projects. Also many a time the best intentions of the government/NGOs and the corporate sector are scuttled due to filmsy reasons such as the historical importance of the area on which the project is to come up or the target users not being of the same castes/sub-castes as the local politician or the land in question is being used by the local community as a public toilet for many years. Reasons such as these contribute to overall increase in costs of the health project and increased suffering of the local population.

Preventative Measures are Also Still Neglected

In general, there has been no pan Indian movement to create hygiene awareness amongst people in rural areas and the urban poor areas of the country. Sporadic episodes have not proved effective so far and in the long run seem to nullify each other. A need for an umbrella organization involving all the concerned sections of the society is, therefore, critically felt.

The authors suggest the following policy initiatives for improvement of the health sector of the country Many experts have tried to identify the ideal health policy which suits both the society and the individual residing within its norms, and they have tried to arrive at the solution in a variety of ways. The authors strongly feel that whatever may be the path or the direction of the health policy/ programme in the country, the health policy needs to include two components, viz., a short-term component and a long-term component.

The Short-term Component: Selective Intervention

The short-term component, should be financed from both public and private health resources, for creating/maintaining the health infrastructural network of hospitals, dispensaries, health centres, training rural health practioners (both paramedical and medical) and stocking them with medicines and medical supplies. Day-to-day financing and stocking of these facilities can be carried out by 'pay-as-per-status' and 'purchase as per requirement' policy so that the ground work for secondary wave of health reforms can be initiated for more synergy-producing interventions. Developing countries are beset with peculiar problems pertaining to financial, and governmental, delays, which make it impractical to confront all health care problems concurrently. It is, therefore, important that priorities be set. The short-term component has to concentrate on the most severe diseases in the society those which result in high mortality, high invalidism, and those that are endemic in nature of the society. Authors believe that it is high time that the government, the society, its representatives and the affected constituency reach an agreement on the health care policy wherein high on agenda should be the efforts to reduce preventible mortality rates especially among the underdeveloped strata of the society. This is easier said than done for this approach would require creating awareness through social marketing, educating the masses, and training a dedicated cadre of workers. The fallout of these efforts would be augmentation of community-oriented basic health services; and initiation of a range of health policy endeavours within the framework of a comprehensive development policy. Which is to say that investments in the health sector should be focused towards establishing facilities comprising good-quality basic services and medicines. Starting with the most vulnerable people, such facilities should be mobilized to concentrate on disease prevention and reduction measures, such as vaccination drives against infectious diseases such as measles, polio, diphtheria, and tetanus. Deaths relating to AIDS and HIV related infections have grown at an alarming proportion in the country. Creating awareness at a war footing with the help of media and dedicated workers calls for a sensitive coordination and substantial funding of various programmes relating to its prevention from the concerned agencies. An interesting aspect of treating the poor is the need for quick acting drugs. The poor cannot afford to sit idle for long as it burdens them with debt. In this regard any short-term health policy intervention should ensure that either the poor are up and about in a relatively short time or have an alternative option of a livelihood, in the absence of which the health network may lose the affected person to quacks.

The Long-term Component

The long-term component of health policy pertains to eliminating the deficiencies in governance. Usually most of the economic and developmental policies are already in place, what is lacking is the implementation and the plugging of existing lacunae. Because people do not in most cases lack awareness, instead there is a lack of implementable solutions to their problems. Execution of all the developmental policies thus are crucial for the development of a hygienic and sustainable habitation for the poor. Such conditions empower people to satisfy their basic requirements of food, sanitation and housing. The impact of such surroundings will then be felt in emergence of disease free and healthy society.

Religion, Spirituality and Sustainable Development

Poverty has been identified as the greatest obstacle in the path of sustainable development of a community. There are many contributory reasons for poverty, viz., political, social, and psychological. Religions across the world also attribute poverty to inequalities of birth. Besides, there are the issues of wealth, its creation, possession, distribution and our attitude towards it. These issues have to be handled and, as far as possible, allayed with the support of the government, the community and individuals. Scholars of various disciplines like economics, politics, history, sociology, religion, theology, ethics and even law, have debated this aspect of humanity for long and each area has generated an independent solution to the problem. Immense confusions and contradictions have compelled them to look for a unified approach. The various religious texts can provide a clue in this direction. Writings of such nature often act as a guide towards a multidimensional approach to development, by pointing to a non-detrimental solution of the developmental problems, and indicating a path which does not hinder the spiritual and materialistic upliftment of the mankind. The belief that the purpose of spirituality and religion is to help us understand and endure the ups and downs of life through empirical wisdom is fast gaining ground. The authors believe that religion teaches us sensitivity, compassion and equanimity at all times.

Ranjit Dugar in his book *Finding Your Spiritual Centre* says : "It (religion) increases our happiness and that of those around us. Spiritual practices open up the boundless energy within us, bail us out of our crises, and insure us against suffering. Thus, religion is the route to take our life to higher and higher standards of moral excellence, and religious doctrines set these standards". This approach is fast gaining ground because results of investing into endless resources and time have not been commensurate with the alleviation of human suffering and shortage. Rich grow richer while the poor are caught in an unending spiral of deprivation. Paradoxically, we witness this in a world that is blessed with abundance to allow every one of us to live a respectable and comfortable life. Then why is there poverty, suffering and hunger in this world? Do government, community and individuals have any moral responsibility in the face of such a situation in our world? What role or approach should religion and spiritual teachers take as they prepare students to face a world full of poverty, hunger and suffering? Is our understanding of religion and spirituality sufficient enough to deal with this challenging social problem?

The answer to puzzle of sustainable development and poverty alleviation lies among the maze of myriad contributory causes of poverty. Let us examine the above questions against these factors.

Political cause of poverty, points towards the moral bankruptcy of individuals chosen to lead the community. An aspect which is manifested in the social and political isolation of communities, exploitation and unequal societal relations, lack of assets, decision making powers and vulnerability to external and internal upheavals in the environment. Inequities of this nature though explained away by economical and financial jargons often disguise human greed, lopsided spiritual convictions and eschewed ethical learning. Principles regarding ethical governance in religions around the world emphasize transferring the poor from their dependence to independence. The Koran, the Holy book of Muslims goes as far as to lay down four basic tenets of ethical governance :

1. *Unity (Tawhid)*: This axiom indicates the vertical dimension of the ethical system. It provides for freedom of action with the view that each individual is viewed as an integral part of the whole.
2. *Equilibrium (Al'Adl wal Ihsan)*: The axiom provides for the horizontal dimension of equity leaving a lot of freedom for policy details, as, for example, striking an appropriate balance between the needs of present and future generations.
3. *Free Will (Ikhtiy'ar)*: Although individual freedom is guided by broad guidelines, and individuals may travel their own paths, careful intellection is required "to interpret-reinterpret that freedom within specific societal contexts, and to suit the needs of changing times".
4. *Responsibility (Fardh)*: The axiom states that although "Responsibility" is voluntary, individuals and society need

to conserve of the public good, since there is a social aspect of every asset owned or managed by private or public entities.

Source: Sirageldin, Ismail (2000), *Elimination of Poverty : Challenges and Islamic Strategies*.

What emerges is the importance of unanimity and balanced approach to development, so that political forces are able to identify the drivers of unification in and among various communities that it has chosen to represent. Authors are of the opinion that the term unification should not be taken on its face value and instead understood in conjunction with human freedom and choices; and also in concurrence with the opportunities that arise from them. It falls on the society to create such an environment and it falls on the individual to behave responsibly. The combination of spiritual development and the human desire for upliftment are thus integral for a dynamic transformation of the political leadership, which in current times has to rise above narrow religious concerns in the spirit of *"Vasudhaiv Kutumbakam"*, which simply put means, the entire humanity is one.

Social cause of poverty, implies that the society as a whole is responsible for its condition, because for long it has avoided a review of its rules of conduct and their application in daily life. A gradual erosion of restraint, friendliness, cooperation, compassion all around us is a far cry from what religion perceived its societies to be, viz., possessing a worldview that is enlightened, a knowledge which is liberal and a conduct which is magnanimous. In other words, spirituality in a society is indicative of its commitment to discipline. The Terapanth sect of Shvetambar Jains founded by Acharya Bhikanji in 1760, established the doctrine of an ideal society on—

1. Equal Distribution,
2. Equanimity,
3. Harmony,
4. Management, and
5. Avoidance of Conflict.

Source: Dugar, Ranjit (2003) *Finding Your Spiritual Centre—The Select Teachings of Jain Sage and Seer Acharya Mahapragya.*

As a part of democracy, the society has to resolve numerous tenacious issues with the help of constructive dialogue. It must be understood that no section of the society commands sole rights of livelihood, beliefs, actions, thoughts, and resources and, therefore, democracy must exist with religion and *vice versa.* A dynamic society means a dynamic and integrated religion. Such a religion is realistic and viable and not trapped in a web of jargon. So long as society does not treat religion as a predicament and compulsion, but as a method to conduct itself; it stands a chance to effect transformation in human ethos. As Indians we need to redirect our energies away from the acts that are proscribed by the religion as amoral, unjust and wrong.

Psychological cause of poverty indicates the need to acquire more and more, even if it is at the expense of the other person in the society. Though livelihood by correct means is often suggested as the appropriate path for acquiring wealth by most of the religions, Buddhism has stated it as one of the factors in the Noble Eightfold Path. It says that the Right livelihood is determined by :

Not by the amount of material wealth it produces, but rather by the well-being it generates. Many livelihoods which produce a surplus of wealth simply cater to desires rather than providing for any true need. For the individual, the objective of livelihood is to acquire the four necessities or requisites of human existence: food, clothing, shelter, and medicine. Again, the acquisition of these four requisites, be it in sufficient amount or in surplus, is not the ultimate objective. The four requisites are merely a foundation upon which efforts to realize higher objectives can be based…The highest kind of person enjoys life on both the mundane and the transcendent planes as follows :

Mundane :

1. Seeking wealth lawfully and honestly,
2. Seeing to one's own needs, and
3. Sharing with others and performing meritorious deeds.

Transcendent :

4. Making use of one's wealth without greed, longing or infatuation, heedful of the dangers and possessed of the insight that sustains spiritual freedom.

Source: Payutto, Ven, P.A (2005) *Buddhist Economics—A Middle Way for the Market Place.*

On a personal level, individuals are advised by the religion not to accumulate wealth beyond a limit where is it. Attachment to wealth is also considered a futile exercise, primarily because religion sees an individual as an instrument through which the humanity as a whole can benefit. It recognizes that due to impoverished state, people may be too preoccupied with the struggle for survival to do anything for their own perfection. Religion, therefore, directs people with means to assist them in acquiring basic requirements. It thus clearly states that as soon as an individual's mental and physical needs are satisfied, he is qualified, and (therefore should be) motivated enough to work for a perfect society. "While wealth as a resource for achieving social good can help create favourable circumstances for realizing individual perfection, ultimately it is mental maturity and wisdom, not wealth, that bring about its realization. Wealth mistreated and abused not only obstructs individual development, but can also be detrimental to the social good"(Payutto, 2005).

From the religious/spiritual perspective, sustainable development is related to the three interrelated aspects of humanity: human beings, society and the policies of the Government. Developmental investments must be in accordance with the whole process and to do that it must have a proper relationship with all three of these areas, and they in turn must be in harmony and mutually supportive. Religion cautions that all sustainable activity must be conducted in a way that it does not harm the interest of the people (by causing a decline in the quality of life) and cause strife in society or imbalance in the physical world. The authors believe that religion holds a special regard for the poor and, therefore, those who are able; should direct their resources towards training, and developing competencies among them so that they (the poor/underprivileged) are able to recognize their own spirituality and are competent enough to critically analyze the choices that they make in the society. At the same time, the society needs to direct its energies towards constantly reassuring the 'have nots' that the involvement of the society in the developmental process is an ongoing exercise and not a knee jerk reaction to certain episodes Integrating the spiritual concepts from all the great religions would thus have a humanizing influence on the process of sustainable development. In the words of Ondari (2001) this would help in turning around ailments like the "troubled and unequal gender relations, the discriminating and isolating social relations, the lack of protection and peace of mind, and all forms of disregard and abuse by the more powerful. Learning the value of hard work; the sacredness of the marriage and the family institution; the sanctity of life and God's special regard and value for all mankind regardless of colour, race, tribe, gender age or creed, is fundamental in transforming most of the cultural practices that cause troubled social and gender relations".

The Hindu philosophy emphasis on '*Karma*' is simply great as it puts emphasis on "action". Rather than being a passive spectator, it is essential that a person strives for excellence in his/her domain of work. It is not about achieving excellence outside the domain of work but it is about being innovative in the every area of work. The debate on destiny is complex and it can go on. Life is a testimony to the fact, that if inclinations are strong, determined action is there, then victory is not far behind. The indicators of success have to be relooked at. Is it about accumulation of personal wealth? Is it about creating an organisation? Is it about creating a name for oneself? The Peace of mind and serenity are objectives which we all value. However, in pursuit of these we get estranged with their surrogates as money, fame and power. A person with contentment has higher degree of peace than running after pursuits which yield more dissatisfaction. A clean mind, limited needs and selfless *Karma* "by excelling at the work one does" with a compassionate mind can transform the work culture. This has to be at the core of any reform process. If each individual becomes conscious of his own duties, then governments, society, organizations will all reform because it is excellence and selflessness that will end all conflicts.

Individual attitude change is instrumental in bringing out the organizational change. It is only sincerity, a noble mind, selfless action, determination and clarity of purpose in life which has to be underlying philosophy in all action – which can transform the languid and passive soul of this country into a dynamic and a progressive nation. The governments must realise that they will have far higher benefits if they are a nation which is vibrant, literate, healthy and self-sufficient. Illiteracy cannot insure immortality. The governments need to rethink about these aspects.

The society needs to be strong and voice its opinion. Societies need to have a voice and must find an opportunity for expression. We deserve what we get if we do not stand up for things in life. We must consciously choose who should govern us. We must be persistent in educating ourselves, fight for our rights. Unless we are demanding we will always compromise. The biggest lesson to be learnt from

Gandhiji's life is the will power to achieve a goal in a peaceful manner. Einstein has, therefore, so aptly put together his views on Gandhiji, "Generations to come will scarcely believe that such a man ever walked in flesh and blood on this earth".

We have a huge reservoir of energy in each one of us. All this energy has to be unleashed and channelised in the right direction. The power is all within us. We need to realise this and unfold this power to achieve the zenith of true selflessness. The big and mighty civilizations have crumbled, the monuments have decayed, the splendour is being mitigated in sands of time, age defies beauty – everything is cyclical and has a limited life. If fundamentally, we recognise this in our lives, we would be so different as human beings in our conduct and our workplace. Work, have a direction and stand for right cause. Identify the cause that is right from the community perspective. Fighting for land, for money, power is all in vain. Resources need to be pooled and be used for bringing in peace and equality.

The firms that operate in a society also need to have a moral responsibility. They are using scarce resources. They need to use them well so that the assets multiply, people get employment, funds can be used for making the life of community better.

Note: Different religions set different limits to wealth accumulation.

REFERENCES

Balaji, V.; Rajamohan, K.G.; Pandy, R. Rajasekara; Senthilkumaran, S. (1999) "The Information Village Experiment in Pondicherry" (online) (cited on 16th December, 2004). Available from <URL:http://itt.nissat.tripod.com/itt20004/infovill.htm>

Caldwell, John C. (1986), "Routes to Low Mortality in Poor Countries". *Population and Development Review* 12:171-220.

Daniel Little, (1990) "Putting the Poor First: Alternative Development Strategies For Third-World Poverty" Conference Paper, SSRC-MacArthur Foundation Fellows in International Peace and Security, Budapest

Dugar, Ranjit (2003), *Finding Your Spiritual Centre—The Select Teachings of Jain Sage and Seer Acharya Mahapragya,* Kolkata, Spiritual Quest.

Dyson, Tim and Mick Moore. (1983), "On Kinship Structure, Female Empowerment, and Demographic Behaviour in India". *Population and Development Review,* 9:35-60.

Heyzer, Noeleen, (1992), "Gender Issues in Anti-Poverty Programs in Asia—

Experiences and Issues", World Bank Institute (online) (cited on 20.9,2004). Available from <URL:http://www-wds.worldbank.org/servlet/WDSContentServer/WDSP/IB/1999/11/19/000094946_99110505342321/Rendered/PDF/multi_page.pdf>.

Mason, Karen, Oppenheim and Smith, Herbert, L. (2003), "Women's Empowerment and Social Context: Results from Five Asian Countries" (online)(cited on 3rd December, 2004). Available from <URL:http//poverty.worldbank.org/files/13323_womens_empowerment.pdf>

Ondari, William, O. (2001) "Poverty and Wealth: A Christian Perspective" *28th International Faith and Learning* Seminar held at Babcock University, Ilishan-Remo, Ogun State, June 17-29, 2001, Nigeria.

Payutto, Ven, P.A (2005) Buddhist Economics: A Middle Way for the Market Place (online)(cited on 7th January, 2005). Available from <URL:http://www.UrbanDharma.com>

Sirageldin, Ismail (2000), "Elimination of Poverty:Challenges and Islamic Strategies" *Keynote Address presented at the 4th International Conference on Islamic Economics and Banking, Islamic Finance: Challenges and Opportunities in the 21st Century.*" August 13–15, 2000. Loughborough University, Loughborough, UK Working Paper 2018.

World Bank. 2001. *Engendering Development—Through Gender Equality in Rights, Resources, and Voice.* Washington, DC and New York, NY: World Bank and Oxford University Press.

Appendix 1

Where Machines Accumulate Man Decays*

J.K. Mehta

Talking to M. Frydman, a Polish engineer, Gandhiji said in 1938, "To use the language of the Bible, what shall it avail a man if he gain the whole world and lose his soul?"[1]

World is on one side, the soul is on the other. Matter is on one side, the spirit is on the other. *Prakriti* is on one side, *Purush* is on the other. Pleasure is on one side, happiness is on the other. You can choose whichever you like: you cannot have both. Look either towards with matter or with the soul. Gandhiji's objective of life is realization of the *self*. It you have to attain this objective you have to choose that mode and pattern of life that would enable you to draw your attention away from matter and focus it on the indwelling soul. Does machinery enable you to focus it on the indwelling soul. Does machinery enable you to focus your attention of the *self* within you?

Gandhiji said on the 8th of September 1920 that "Europe is today only nominally Christian. It is really worshipping Mammon. "It is easier for a camel to pass through the eye of a needle than for a rich man to enter into the Kingdom of heaven. Thus really spoke Jesus Christ. His so-called followers measure their moral progress by their material possession."[2]

Why is it difficult for a rich man to enter into the Kingdom of heaven? Because a rich man's mind is focused on all the material

---

* Lecture delivered by the well-known Gandhian economist Prof. J.K. Mehta of Allahabad University on *Man and Machine* reproduced here for the benefit of the readers.

means of living. And machinery helps us to satisfy our craving for such material means. Let us repeat that Gandhiji was against machinery only because its use made it easy for man to adopt ways that would make it difficult for him to realise his *self*. It is not the machine, of course, but its use that led to man's degeneration. Is it at all possible to make such an use of a machine as would not prevent him from realizing his *self*?

Let us see what a machine is. A machine is a contrivance in which the final, operating part is driven by mechanical, electrical or atomic power, in short, it is driven by non-human power. A tool, on the other hand, is a device that needs for its operation human power. A simple knife, a simple screw driver, an ordinary hammer, a hand-operated plane are examples of tools. A sewing machine a cycle, an electric fan and such things are examples of machines. In the case of these machines the final moving parts (the needle in the case of a sewing machine, wheels in the case of a cycle and the blades in the case of a fan) are supplied with some mechanical or electric power. Between the final moving part and the human agent there is some mechanical or electrical go-between.

Gandhiji often said that he was not against the use of tools, nor against the use of simple machines or even sophisticated machinery if they did not exploit the poor people or throw them out of employment. When Gandhiji makes such statements where he is not seen to be unequivocally opposed to the use of machinery, it becomes necessary for us to go into the question a little more carefully. The use of every contrivance, or every kind of use of any given contrivance, has to be judged in the light of its contribution to our progress towards the realization of the *self*. Every means has to be used as a means to a given end. Gandhiji's end was realization of *self*. In what way and to what extent does any particular aid to production help man to attain this objective?

To speak in the language of economics, a tool or a machine is a capital-good. It is the result of the application of labour to some gift of Nature. It represents, in a way, congealed labour. Its use increases the efficiency of labour in an act of production. But whether it is a simple tool or a complicated piece of machinery, a capital-good is the result of what we might call man's enslavement of Nature's gifts. It is by enslaving Nature or by harnessing natural forces that we are able to increase the effectiveness of labour as a factor of production. Just as we can use the services of human slaves to satisfy our wants, so also we can use the services of Natural slaves to satisfy more easily our wants. We are opposed to slavery which has more or less been rooted out from our society. It requires our going a step further in the same direction to feel opposed to the slavery of Nature's forces. It is bound to sound ridiculous to some but not to all that we should be compassionate to Nature in the same way in which we should be compassionate to human beings.

Call it compassion or what you like; the fact is that man has made himself believe that all that exists in the Universe has been created by God for his exclusive use. He thinks he can enslave all forms of animal life and rob Nature of all her wealth for his own good without incurring His wrath. Gandhiji was conscious of this and that is why he wanted that we should be as self sufficient as possible. There are other reasons for which he preached self-sufficiency; but at the bottom of all of them there was the philosophical or moral reason – it is a sin to exploit others, it is a sin also to use Nature as our slave.

That is the depth charge on machinery. All of us cannot think so deeply and so we have to think a little superficially and see how the use of machinery is inconsistent with the objective of life that was so dear to Gandhiji. It is not difficult to misunderstand Gandhiji; he is not always consistent in his statements. Nor did he ever care to be consistent. He cared more for truth as it revealed itself to him from time to time than for consistency. Nor did Gandhiji care for formal logic and he did not depend on reasoning so much as on his intuition. All this is responsible for causing a good deal of misunderstanding of his views. At times he says he is against the use of machinery, at other times he says he would welcome even sophisticated machinery under certain conditions. It is necessary for this reason to make always an analytical study of Gandhiji's thoughts. And we have been attempting to do that in our lectures.

So let us proceed analytically to study the problem that is facing us here. Let us bear the fact in mind that a tool may be called a simple machine and a machine a complicated tool. In both the cases some use of natural power or nature's material is involved. We have observed that machines and tools are capital-goods that increase the efficiency of labour as a factor of production.

Let us start with the simple case of production by an individual so that all social problems can be ignored. An individual as a producer has to use some kind of capital-good except in the simplest and most primitive acts of satisfaction of want. When a hungry man goes to a tree and plucks fruits for his consumption he performs an act of production without having to use any kind of capital-good. There is no use of a machine nor even of a tool.

In all other cases man makes use of some tools and machines in order to satisfy his wants. It might appear that when social relations are not involved there is no harm when a man uses capital-goods. He does not then exploit the labour of other people and does nothing else that could interfere with his efforts to attain his ultimate objective, namely, the realization of his *self.* That, however, is not quite correct. With the use of capital-goods a man is able to increase his productive efficiency and thereby make his life more comfortable and, in some cases, more luxurious also. With every increase in his efficiency as a producer a man, therefore, begins to concentrate his attention more and more on the means to satisfy his material wants than on the ways of satisfying his spiritual wants. His mind begins to dwell constantly on worldly things and it becomes difficult for him to remain detached from them. What is necessary for the realization of the *self* is to cultivate an attitude of indifference to worldly things and, while doing all that is necessary to keep body and soul together, think constantly of the indwelling *self* within us. By nature man is greedy and ease-loving and, therefore, the use of machines tends to make him more ease-loving and greedy and thus comes in the way of his endeavour to realize his *self.* The more automatic and sophisticated the machinery the greater is the ease with which man can satisfy his wants and greater, therefore, is the hypnotic influence that it exercises on his mind. His mind is tempted to dwell more and more on matter than on the spirit. He tends to gain the world more and more and lose more and more his soul, to speak the language of the Bible.

Our mind has a natural tendency to look into the external Universe and dwell on the things that it find there. The greater the ease with which he can satisfy his wants, the greater the amenities of life that enrich his living, the greater is the force with which a man's mind gets attached to the material means of satisfying wants. If our mind has to turn its direction inwards and seek the knowledge of the *self,* if it has to become easier for it to detach itself from matter and attach itself to the soul, the glamour of the external Universe has to be dimmed. It is the use of machinery that is responsible for making our mode of living glamorous. The things that machinery helps us to produce place, temptations in our way that draw us farther and farther away from the soul. That is why we have said that where machines accumulate man decays. And so did Oliver Goldsmith say in *The Deserted Village,* "Ill fares the land, to hastening ills a prey, Where wealth accumulates and men decay."[3] And Henry David Thoreau says in *Walden*: *Conclusion*, "Superfluous wealth can buy superfluities only. Money is not required to buy one necessary of the soul."[4] Here wealth is said to exercise a deadening effect on the spiritual health of man, both by Goldsmith and Thoreau. And the use of machinery increases our ability to earn and accumulate wealth.

There is one characteristic of production in the modern world: it is machine-based and it functions through the medium and intervention of money. The former increases material wealth and the latter enables man to acquire and store it up. In the absence of money wealth would have had to be consumed at once or given away to others. But where there is money, one can convert the excess of wealth over the immediate needs of consumption, into money and keeps one's control over it. This is the greatest evil of the otherwise useful institution of money.

Be that what it may, Truth takes every opportunity to reveal itself. It peeps through every hole in the physical personality of man. That is why whether it is a saint, a poet, a philosopher or a musician, he in his moments of calm contemplation realizes that the fact that one cannot have both worldly goods and spiritual calm, the supreme Bliss of the soul. There are only two things; (1) there is the immortal spirit, the Infinite that has no beginning and no end, the soul that is existence, consciousness and bliss absolute, and (2) there is matter, the finite entity that has beginning and an end, the phenomenal objects of the Universe that have transient existence, that live, as it were, on borrowed consciousness and happiness reflected from the soul. One can have either but not both of these. Our mind can think of and dwell on the external material world or the internal spiritual one. Since it is the final objective of all beings to realize the *self* attempt must be made to detach ourselves from matter and attach ourselves to the soul. The one thing that stands in the way of such an attempt is the glamour of the external world. The use of machinery enhances this glamour.

Now let Gandhiji speak. "You cannot serve both God and Mammon. God the compassionate and the merciful, Tolerance incarnate, allows Mammon to have his nine days' wonder. But I say to you.......fly from that self-destroying but destructive show of

Mammon."[5] (*Young India*, 8th December, 1927). That is why we have said that you can have either but not both of them. You cannot at the same time walk in two opposite directions.

Again Gandhiji says, "Civilization in the real sense of the term, consists not in the multiplication, but in the deliberate and voluntary reduction of wants. This alone promotes real happiness and contentment and increases the capacity for service".[6] (From *Yervada Mandir*, Chapter 6)

That is true; we have already seen how machinery increases wealth and the glamour that goes with it leads to increase of wants. Happiness and contentment are to be found in simple living with reduced wants. It paves the way for the realization of the *self.*

Condemning the use of machinery Gandhiji says "do want growth, I do want self-determination, I do want freedom, but I want all these for the soul. I doubt if the steel age is an advance upon the flint age. I am indifferent. It is the evolution of the soul to which the intellect and all our faculties have to be devoted".[7] (*Young India*, 13th October 1921). Gandhiji was not in reality indifferent. He was openly opposed to the achievements of the steel age. And why? As always, because that stands in the way of our efforts to realise the *self.* That is our final objective and everything that helps us in attaining that objective has Gandhiji's approval.

Gandhiji emphasizes the same fact when he says, "High thinking is inconsistent with complicated material life based on high speed imposed on us by Mammon worship. All the graces of life are possible only when we learn the art of living nobly".[8] (*Harijan*, 1st September, 1946). In simple words it means that machinery makes life complicated. Complications and simplicity cannot go together. It is simplicity of life in which all the graces are to be found. It is in such a life that it is easy to make a successful attempt at the realization of the *self.*

Gandhiji said in 1936 that "A certain degree of physical harmony and comfort is necessary, but above a certain level, it becomes a hindrance instead of help. Therefore, the ideal of creating an unlimited number of wants and satisfying them seems to be a delusion and a snare. The satisfaction of one's physical needs, even the intellectual needs of one's narrow self, must meet at a certain point a dead stop, before it degenerates into physical and intellectual voluptuousness".[9]

Read in between these lines, read under and behind these lines and you will realize that Gandhiji wants that the sleeping *atma* of man must be aroused from its sleep and made to wake up to find that it is *atma* and not matter. We know that the *self* out of ignorance identifies itself with the body and mind and the ultimate objective consists in making it realize itself. We have said a number of times that one does not know why God exercised His *mayasakti* to throw this veil of ignorance on the *atma*. But we know that God Himself wants that this ignorance should go and the soul must reestablish itself on its throne.

It is for this reason that all great men have said that the less the material means of living are there, the easier it is for us to realize our *self.* This is the most fundamental point against the use of machinery which, while supplying us the needed necessities of life, tends to cross the limit and leads, as Gandhiji said, to physical and intellectual voluptuousness.

Gandhiji is difficult to understand and often easy to misunderstand. That is because he does not care for logic nor for consistency. He used to say that he did not care for consistency and he said and did what appeared to him to be correct. That is why he said one thing today and another the next day. He advised us to accept as correct (to the extent that he could make a correct statement) his latest utterances. Many of us, therefore, feel confused when we read his views on machinery. We shall have, therefore, to proceed cautiously while studying his views on machinery.

In 1924 Ramachandran, a student of Shantiniketan was introduced to Gandhiji by C.F. Andrews. Ramachandran asked Gandhiji, "Then you are fighting not against machinery as such, but against its abuses which are so much in evidence today?" Gandhiji replied, "I would unhesitatingly say, yes, but I would add that the scientific truths and discoveries should first of all cease to be the mere instruments of greed. Then the labourers will not be overworked and the machinery instead of becoming a hindrance will be a help. I am aiming not at eradication of all machinery, but limitation".[10] Now here is the point I have been trying to make out. Gandhiji says that he is *not aiming* at eradication of all machinery. As a matter of fact, he *was* really aiming at its complete eradication. If we remember his objective of life we shall have no difficulty in understanding the fact that he was *aiming* at the total eradication of machinery, though he realized that we could not do that all at once. We should minimize its use, limit its use, but our final objective is to give up its use altogether.

When Gandhiji, illustrating his point, said that he was not against the use of such machines as the Singer Sewing machine, Ramachandran told him, "But in that case, there would have to be a factory for making these Singer Sewing Machines, and it would have to contain power-driven machinery of ordinary type." Gandhiji replied by saying, "Yes, but I am socialist enough to say that such factories should be nationalized or state-controlled. They ought only to be working under the most attractive and ideal conditions, not for profit, but for the benefit of humanity, love taking the place of greed as the motive....... This mad rush for wealth must cease and the labourer must be assured not only of a lean wage but a daily task that is not a mere drudgery".[11]

Does it not appear from this conversation that Gandhiji was in a way giving away his point? He would allow the use of power-driven machinery and factories. But Gandhiji in the heart of his heart knew that so long as there was machinery in use there would be impediments in the way of our realizing our *self*. At some places Gandhiji has said that he would not have industrialization even in socialism or communism. There are apparent contradictions in Gandhiji's utterance. But they are bound to be there in the case of a man who was *experimenting* with Truth.

Gandhiji would have no objection to the use of simple tools and simple machines, as he said, if they did not lead to exploitation of labour. All machines should be used, if they could, for the benefit of humanity. But for the manufacture of simple machines we would need complicated and power-driven machinery. We would have to tolerate or even welcome them. Complicated, sophisticated, power-driven machines are required to produce simple machines. The process appears to be reversed: complicated machines first and simple machines afterwards. Simple machines can be produced only by complicated machines. But how are complicated machines produced? Are they produced by still more complicated machines? If so, where does the process end? Any way, if complicated machines constitute the starting point, how did we get them in the beginning?

This question suggests the answer. In all cases we begin with simplest tools, made of stone or other metals. These are produced as a result of the application of human labour to Nature given gifts. With the help of such simple tools we make some complicated tools and with their help, again, we make small machines. Using small machines we make big, complicated machines. Once big machines are produced, we use them to make small machines. Small machines are then used to produce still smaller machines and ultimately tools. From there onwards, it is no more necessary to make tools with the application of our physical labour to the gifts of Nature.

If, therefore, the use of big, sophisticated machines is to be avoided we should stop at the level of simplest tools that can be made with our hands and not pass beyond that stage even to produce machines like the sewing machine that could reduce the imagined drudgery of human labour. Gandhiji's ideal, whether he mentions it or not, was to use the simplest possible tools so that man could make the greatest possible use of the arms that God has given him mercifully and thankfully to use the gifts of Nature. Let us repeat that there are apparently many contradictions in Gandhiji's statements. At times he draws our attention to the ideal, at other times he tells us how we can make the best a bad job. He would prefer the bullock cart to the railways; but when the railways are there and our economy has patterned itself on the infrastructure of the railway system then we have to continue to use the railways making their ownership and operation as little harmful to the interests of society as possible.

Let us repeat that what is consistent with Gandhiji's objective of life is not to go beyond the stage of simple tools that man can make with his hands. Animals (except man, who is also an animal) do not make even tools. And if they do not use even tools could we say that they must be finding it easier to attain the ultimate objective of realizaton of *self*? This question can be answered but it would take us into the by-lane into which we would not like to go here.

Does God want us to use simple machines or tools? It has been said by some biologists that man's progress on the material plane and the modern civilization which is its outcome have been made possible by the peculiar anatomical structure of man's hand. A man has a thumb and four other fingers and the thumb is so structured that he can hold things in his hand. Other animals (with the exception of apes) cannot hold sticks and other objects in their hands. It is because man can use his hands to hold sticks, rods and such other objects that he has been able to make tools and implements. By the way, it is also other animals far stronger than himself.

If the human anatomy is such as would make it possible for a man to make tools and then use them to make machines, does it follow that God wants that man should make and use machinery, not only simple machines, but very sophisticated power driven ones? No, God is not responsible for the human anatomy that appears to have been designed for the machine age. It is the ignorant soul that can be held responsible for it. By His *mayasakti* God has made the soul so ignorant that it identifies itself with matter and tries to glorify itself by making the best possible use of its power to sink deeper and deeper into the material mould into which it is cast. Out of ignorance, the soul of man is seeking its fulfilment in the sensual life where pleasure rather than happiness is all that can be had. This is the process of evolution in *kaliyuga*.

So to go back to our main point, let us repeat that whatever Gandhiji might have said from time to time, what is consistent with his objective of life is the least possible use of material means of satisfying wants. In fact, in the final analysis there would be no tools and no machines in Gandhij's ideal society. Even a simple tool used by a man for the satisfaction of his own wants, where there is no possibility at all of exploiting anybody, is an impediment in the way of the attainment of the final objective of life. And let it be repeated once again that the use of tools and machines is for Gandhiji a necessary evil. Necessary for the reason that the shackles of the machine age. It is an evil, that we are forced by circumstances to tolerate.

What we have said is borne out by the conversation that Gandhiji had with a Polish engineer M. Frydman in 1938. He asked Gandhiji "Is no compromise with the industrialization possible without imperilling the ideal of self-sufficient villages"? Gandhiji replied by saying, "Oh, yes, railways are there I do not avoid them I hate motor-cars, but I make use of them willy-nilly all the same. Again I dislike fountain-pens, but just now I am making use of one, though I carry a reed pen about in my box. Compromise comes in at every step, but one must realize that it is a compromise, and keep the final goal constantly in front of the mind's eye".[12]

So let us say again that Gandhiji was totally opposed to the use of machines and even tools and where they existed and were in use he said we are forced by circumstances to tolerate them, but we must "keep the final goal constantly in front of the mind's eye".[13]

However, let us remember that there could be no objection to the use of machines if we could do our work without any mental attachment to them. A rich man is not rich, or is not sinfully rich, if we does not attach himself to his wealth. That is why Gandhiji wanted that all rich people should behave as if they were trustees of the poor. So, while using machines if we could keep our mind detached from them, i.e., in other words, if we could use them without having the desire to enjoy the fruits of such use for our own selves, there could be no objection to machinery. But man being what he, is such a detached attitude to the use of machines is not possible. When a man becomes a saint he develops that attitude, but when men become saints they would not produce machines.

If you want to know Gandhij's views regarding any social, political or economic matter you have only to ask yourself whether such and such a thing would help us to realize our *self.* If it helps, Gandhiji must have said that. If machinery helps us to realize the *self* Gandhiji would approve its use, if it is an impediment in the way of self-realization Gandhiji would not have anything to do with it. That is the easy way of knowing Gandhiji's views. It is not necessary to scan through the pages of the numerous books that have been written on Gandhiji. That does not mean that nothing can be gained by reading these books. If you have to make an analytical study of Gandhian Thought, as we are making, you must carefully read all that Gandhiji has written.

Now let us note what Gandhiji said in Poona when the Education Ministers from the Congress provinces met in conference in July 1946. "Ours has been described as the machine age, because the machine dominates our economy. What is a machine? One may ask. In a sense, man is the most wonderful machine in creation. It can neither be duplicated nor copied."[13] Gandhiji, however, used the word machine in its ordinary sense in which it tended to displace the human or animal labour instead of supplementing it or merely increasing its efficiency. That was the first differential characteristic of the machine. The second characteristic was that there was no limit to its growth or its evolution. That could not be said of human labour. There was a limit beyond which its capacity or mechanical efficiency could not go. Out of this circumstance arose the third characteristic of the machine. It seemed to be possessed of a will or genius of its own. Machine was antagonistic to man's labour. Thus, it tended to displace man, one machine doing the work of a hundred men, if not a thousand, who went to swell the army of the unemployed, not because it was desirable, but because that was its law.[14] (Quoted from *Mahatma*, by D.G. Tendulkar, Vol. 7)

We would like to note carefully the second characteristic of machine that Gandhiji has mentioned here. He sys, there is no limit to its growth or its evolution. Certainly, machines tend to grow and evolve. That is why we said earlier that once we have made a tool it tends to evolve into a machine and a machine into a complicated bigger machine and so on, till a time comes when these big power-driven machines begin to produce simple machines and tools, no more needing the use of human labour. That is how evolution on the material plane goes along with human evolution. Man's mind gets sunk deeper and deeper into the sea of materialism. It becomes increasingly difficult to extricate the mind from the clutches of materialism.

Thus, the use of simple tools if unrestricted, would lead to the use of machines of ever increasing complexity that would retard man's progress on the spiritual plane. But this is *kaliyuga* in which machines and all kinds of mechanical and electrical device tend to make man as loving animal whose mind is turned from the spirit to the matter. The ignorant soul of man under the influence of *maya* becomes more and more ignorant and gets confirmed in its delusion that it is matter—its identification with matter gets more and more strengthened.

In 1946 Gandhiji wrote in the *Harijan* of 15th July, "if, there is no room for machines that would displace human labour, that would concentrate power in a few hands. Labour has its unique place in a cultured human family. Every machine that helps every individual has a place. But, I must confess that I have never sat down to think out what that machine can be. I have thought of Singer's Sewing machine. But even that is perfunctory. I do not need it to fill in my picture".[15]

So here we find that Gandhiji expresses his doubt about the usefulness of even such a machine as a Sewing machine, which he had on other occasions accepted as an innocent and useful adjunct to human labour. That proves our contention; even a simple tool, leave aside a simple machine, is an hindrance to the development of our spiritual personality. Gandhiji is right when he says that he had thought of Singer's Sewing machine as one that would help every individual but he had not sat down to think out the problem and that his previous views on the sewing machine were perfunctory.

So let us say that where machines accumulate *man* decays. We have deliberately used the word man in the singular to emphasize the fact that machines are an impediment to the development of human personality. Even Robinson Crusoe, who has no relationship with any other human being, would find the use of machines detrimental to his spiritual well-being. Crusoe as a *man* would decay. But when we take up the case of men who live and work together, where there is social organization, more complications arise. In that case decadence of human beings takes place. The user of machines drags down along with himself those whose services he employs to operate them.

When we read through Gandhiji's writings we find that his main and often, it seems, the only objection to the use of machines is that it leads to the concentration of power in the hands of a few people and the unemployment of a large number of workers. We shall quote passages from his writing in support of this statement. But we might ask a pertinent question here. Did Gandhiji want that Mr. *A* should not use machines because that would cause hardship to other people? Did he think that machines should not be used by Mr. *A* so that the well-being of the people other than himself may be safeguarded? Since Gandhiji's avowed objective was realization of the *self* did he think that the use of machines would make it difficult for other people to attain this objective? Or did he think that when Mr. *A* adopts a technique of production that leads to concentration of power in the hands of a few and results in the unemployment of many, Mr. *A*'s own interests suffer as he would find it difficult to attain his objective of realization of the *self*?

We could go a step farther and ask whether Gandhiji was concerned about the well-being of the users of machines, or of those whose services were hired to operate them or was he primarily concerned about his own well-being? Was he not denouncing the use of machines because by so doing, he believed, he would make it easier for himself to realize his *self*? His objective, first and foremost, was to realize the *self*, he wanted to see God face to face. Asking people not to mechanize production was for him a way of achieving his objective. No, that was not Gandhiji's first and foremost objective, it was his first and last objective. He wanted to serve himself and if he served other people that was because he felt that that was the way of serving his own interest. Shall we then call Gandhiji selfish? We may with impunity call him selfish, but let us remember, everyone of us is selfish. What we want is that our action should be guided by enlightened selfishness. As a matter of fact, there was enlightened selfishness in Gandhiji's thoughts, words and deeds. God wants us to be selfish, to be really selfish; let enlightened selfishness determine all your actions that is what God would say. As a matter of fact we who are selfish are not in reality selfish. We seem to seek our own salvation but we are shortsighted people and we do not know what path to traverse in order to reach that goal. So, in the final analysis, we are not selfish people; only people like Gandhiji are selfish.

Should we then say that selfishness is a virtue? Yes, we should say that, remembering that true selfishness, enlightened selfishness, is a virtue. It is a virtue to seek our own salvation. One might legitimately enquire whether it is a virtue to seek our salvation if that makes it difficult for others to attain *moksha*?

The answer is in the negative. If our enlightened selfishness comes in the way of other people who are wanting and striving to realize the *self*, it cannot be considered as a virtue. But the fact is that such a thing cannot happen: what is in our interest (our enlightened selfishness) must also be in the interest of other people. We cannot get happiness at the expense of the happiness of other people. Our selfishness, if it is of an enlightened type, cannot in any way stand in the way of other people seeking salvation. That is ensured by the fact that while the bodies and minds of people are different, their souls are the same. Anything that we can do to liberate our soul will naturally lead to the liberation of other people's souls.

Let us in this connection recall what has been said in the *Upanishads*. It is said there, verily, the whole existence is Brahman. This Self is the Brahman. That thou art and I am the Brahman.

Let us never forget that soul is infinite, it is not material and, therefore, it is omnipresent. It is just One, it has no parts and it cannot be said to be here and there. It just *is*. If I do anything that helps the liberation of *my* soul it automatically helps the liberation of the souls of other people. As a matter of fact, this statement is made for the sake of convenience of expression. Otherwise, there are not souls but just (one) soul.

To go back to our main point, Gandhiji objected to the use of machinery in an economy on the ground that it would lead to concentration of power in the hands of a few and result in the unemployment of many. Where, he often said, machines could be so devised and used as neither to cause unemployment nor leads to the concentration of power in the hands of a few people, he would welcome their use. As a matter of fact, for the reasons explained earlier, it is not possible, as Gandhiji realized himself, to avoid the evils of the machine system even where a society is organized on socialistic lines. When Jawahar Lal Nehru told Gandhiji that the use of machinery would be innocuous in socialism he replied by saying that there were built-in evils in industrialization which could not be avoided whatever the type of economy we had.

Superfluously considered it would appear that Gandhiji's point of view was rather shallow. He was opposed, it would seem, to industrialization only because it would oppose, it would seem, to industrialization only because it would lead to unemployment and exploitation of many by a few. As a matter of fact, Gandhiji believed that so long as people were unemployed and exploited and he made no effort to help them out of their difficulties he could not make any spiritual progress, i.e., he could not proceed towards the realization of the *self.*

In witness let us quote Gandhiji, "A factory employs a few hundred and renders thousands unemployed I may produce tons of oil from an oil mill, but also drive thousands of oilmen out of employment. I call this destructive energy, whereas production by the labour of millions of hands is constructive and conducive to the common good. Mass production through power-driven machinery, even when state-owned, will be of no avail". Mark the words here; "even when state-owned, will be of no avail".[16]

Further Gandhiji said, "God created man to eat his bread in the sweat of his brow, and I dread the prospect of our being able to produce all that we want, including our foodstuffs, out of a conjurer's hat".[17] (*Harijan*, 16th May, 1936)

But we today seem to be wedded to the policy in which everything is attempted to be produced out of a conjurer's hat. It is a sad (deplorable) commentary on the aftermath of our Independence that Gandhiji's views have been forgotten and his advice neglected with impunity. Only his name is remembered. We go on repeating his name without showing any respect for his thoughts, words and deeds. Had Gandhiji been living today, he might perhaps have said, "take not my name in vain".

Gandhiji said, "One has to go only a step further to understand that the result would be the same as between land-made goods and those made by power-driven machinery. W are seeing the process going on before our eyes. Little flour mills are ousting the *chakki*, oil mills the village *ghani*, rice mills the village *dhenki*, sugar mills the village *gur*-pans, etc. This displacement of village labour is impoverishing the villages and enriching the moneyed men. If the process continues sufficiently long, the villagers will be destroyed without any further effort. No Chengis Khan could devise a more ingenious or more profitable method of destroying these villages".[18] (*Harijan*, 20th June, 1936)

What is our government doing today? They are proving themselves more ingenious than Changis Khan. Industrialization is proceeding or rather has proceeded at a galloping rate despite the lip sympathy that we have been showing for Gandhiji and his views. Hypocrisy rather than Truth seems to be triumphing in our land. And what fruits are we reaping? What has planning and what has industrialization and what has our public sector done for our country? If Gandhiji were to come to life again and stand before us, with what face would we greet him? Thank God, we are safe; Gandhiji will not come to life again.

Gandhiji believed strongly that industrialization would lead to concentration of power in the hands of a few. It would lead also to centralization of production. He wrote in the *Harijan* of 30th December, 1974, "I suggest that if India is to evolve along non-violent lines, it will have to decentralize many things. Centralization cannot be sustained and defended without adequate force. Simple homes from which there is nothing to take away require no policing. The palaces of the rich must have strong guard to protect them against dacoity; So must huge factories. Rurally organized India will run less risk of foreign invasion than urbanized India, well equipped with military, naval and air forces".[19]

Two facts here need to be noted. First, Gandhiji wanted decentralization, believing that in the case of centralization (even in socialism) there is concentration of power. He knew that power tends to corrupt and absolute power corrupts absolutely. These were the words of Lord Acton who further said that great men are almost always bad men. If power has to be concentrated let it be concentrated in the hands of each individual. Let an individual be his own master and his own slave. Gandhiji's ideal was self-sufficient individual. A man is a small-scale replica of the entire Universe. The cosmos as *Vedanta* says, is reflected in the individual. There is an internal universe that is within an individual just as there is an external universe that lies outside him. The more one pays attention to this internal universe the closer one gets to the indwelling soul of man.

The second point that we have to note here is that Gandhiji was extolling poverty. Not only is a poor man himself nearer God, he does not tempt others to recede from God. Did not Gandhiji say as Christ said, "It is easier for a camel to go through the eye of a needle than for a rich man to enter into the kingdom of God". Industrialization tends to make some people rich at the expense of others. It creates a class of rich people who would find it hard to enter the kingdom of heaven. It creates another class of poor people; their poverty would be a boon to them but they would not have freedom that should accompany.

Man has always sought power, the power to plunder material wealth from the external universe to fatten his ego. What he needs to do is to exercise his power over his own self to weaken his ego. Let him remember that the power he has been trying to gain and to exercise belongs to God. Let us quote from Psalm (62:10) "If riches increase, set not your heart upon them. God hath spoke once; twice have I heard this: that power belongeth unto God".

The greater the power the more dangerous the abuse. This is what Edmund Burke said. Machinery helps a man to become rich. It helps him to acquire power. And with this power he is tempted to abuse his position. Religious men and politicians have alike spoken against power.

And hear now what Benjamin Disraeli has said, "I repeat....that all power is a trust – that we are accountable for its exercise – that from the people, and for the people, all springs, and all must exist".[20] Power belongs to God. Truly has that been said in the Psalms. When man begins to use machines he robs Nature of its wealth and attempts, often without his knowing it, to rob God of His power. Gandhiji knew all this and he wanted to remove all the temptations in the way of man's efforts to weaken his ego.

And let us end this theme with a quotation from Thomas Gray's *Elegy*. "The boast of heraldry, the pomp of power / And all that beauty, all that wealth ever gave / Awaits alike the inevitable hour / The paths of glory lead but to the grave".[21] The words that sprang from the lips of Gandhiji are the same that were uttered by poets, politicians and prophets. Truth is One, it is immutable. It has no multiple interpretations.

Gandhiiji said, "And surely modern civilization is not millennia old. We can almost give its birth an exact date. If I could do it, I would most assuredly destroy or radically change much that goes under the name of modern civilization. But that is an old story of life. The attempt is undoubtedly there. Its success depends upon God. But the attempt to revive and encourage the remunerative village industries is not part of such an attempt, except in so far as every one of my activities, including the propagation of non-violence, can be described as such an attempt".[22]

Gandhiji would destroy or radically change all that goes under the name of modern civilization. This is *kaliyuga* and modern civilization means the *kaliyuga* civilization. Today, art, music, literature, dress, everything is vulgar from our point of view. So Gandhiji would pronounce modern methods of production also as vulgar. Through truth, through non-violence, through the trusteeship of the rich, through the non-use of machinery and through his own fasts, Gandhiji attempted to destroy modern civilization.

Gandhiji, as I have said earlier, can be misunderstood. That is because he did not care for logic or for consistency. For instance, he wrote in *Harijan* of 30th November, 1934, "I have no partiality for return to the primitive method of grinding and husking for the sake of them. I suggest the return, because there is no other way of giving employment to the millions of villagers who are living in idleness. In my opinion, village uplift is impossible, unless we solve the pressing economic distress".[23] The primitive methods of production would provide employment to idle hands. That is true; but Gandhiji, as is evident from a close study of his writings, would gave the people adopt the primitive methods of production even when there is no idleness. That might *enforce* simple way of living on the people, but that is precisely what Gandhiji wanted. Production-methods that did not involve the use of machines has to be welcomed for their own sake too.

Robert Browning said, "Grow old along with me / The best is yet to be". Gandhiji can be imagined as saying "Grow old along with me, / The Truth has yet to be told." Gandhiji was experimenting with Truth. Whatever he felt he said and when he felt that he was wrong he acknowledged his mistake and offered his revised views on all matters. That is why we have said that he could be imagined and saying "Grow old along with me, / The Truth has yet to be told". Was not Gandhiji simple and unassuming like a child? He was a child all his life. The child is father of the man. He was the father of all of us.

An American friend asked Gandhiji, "do you feel that mass production will raise the standard of living of the people"? Gandhiji replied, "I do not believe in it at all. There is a tremendous fallacy behind Mr. Ford's reasoning. Without simultaneous distribution on an equally mass scale, the production can result only in a great world tragedy. Take Mr. Ford's cars. The saturation point is bound to be reached soon or late. Beyond that point the production of cars cannot be pushed. What will happen then? But it can be definitely shown that mass production carries within it has its own limitations".[24]

We do not quite know whether the point of saturation has come. But we know that this mad rush for multiplication of our wants and the pampering of them at any cost has resulted in many social evils. Are the people of the countries where there is mass production of articles of comfort and luxury really happy? As a matter of fact they find that the wealth that they are producing lies heavy on their head. They realize that they are caught in the labyrinth of wealth where they find all that man could desire except mental peace. W shall not dilate upon this point. Those who are familiar with the conditions of life in the advanced countries of the world will readily agree with us: we have wealth without peace. It is in that sense, if not in any other, that the saturation point has been reached. Let Gandhiji be understood in this extended sense of the word saturation. We have to make an analytical study of Gandhiji's thoughts and words.

What about Mr. Ford's plan of decentralization of industry by the use of electric power conveyed on wires to the remotes corner as possible remedy where there would be hundreds and thousands of small, neat, smokeless villages dotted with factories, run by village communities? The American friend asked Gandhiji if that would meet his objection. Gandhiji said that that would not meet his objection, "because while it is true that you will be producing things in innumerable areas, the power will come from one selected centre. That in the end, I think, would be found to be disastrous. It would place such a limitless power in one human agency that I dread to think of it".[25]

Power belongs to God, let us remember that. We have stated the views of some people who have emphasized the fact that power corrupts. We shall, therefore, not say more about it here but simply note what P.B. Shelley has said, "Power, like a desolating pestilence, Pollutes whatever it touches".[26] (*Queen Mab*).

Where there is machine there is power, the devilish power. You cannot purge machinery of the power that it gives to human beings by replacing capitalism by socialism or communism. When his American friend asked Gandhiji, referring to Russia, whether it was not a country that had developed mass production without exploiting in Gandhiji's sense, the less industrialized nations, he replied "You want me to express opinion of State-controlled industry, i.e., an economic order in which both production and distribution are controlled and regulated by the State as is being done in Soviet Russia. Well, it is a new experiment. How far it will succeed ultimately I do not know. If it were not based on force I would dote on it. But today since it is based on force, I do not know how far and where it will take us".[27]

You will see that Gandhiji believe that in all cases machinery and power or force will go together. Today it is based on force he said. Why should we bank on socialism and communism when we know that what sustains them is *force* and *power*? Is not in effect socialism or communism the epitome of centralization? Wherever there is centralization, which Gandhiji did not like, there is concentration of power; it matters not in whose hands that power is concentrated. Let us recall the eternal truth, namely, that power belongs to God and it is in His hands only that is should be concentrated. When there is complete decentralization where is the power? It is in the hands of individuals. Each man has power over himself. Power, in that case, is purged of its venom. Machinery and decentralization cannot go together. They cannot live under the same roof. Simple tools and decentralization can take shelter under one and the same roof. But while there may be no concentration of power where simple tools are used, there are other evils attached to their use. We have explained that earlier and so we shall not repeat the explanation. However, let us repeat once again, what really bears repetition, that Gandhiji liked all those things that help us to realize our *self* and disliked those stand in the way of *self*-realization.

The Western world has at least come to realize that too much of a good thing is bad. The American friend referred to above, told Gandhiji, "This is a particularly appropriate moment, it seems to me, for you to be visiting London, because, apart from the political questions, it seems to me to be a time when the Western world is disillusioned in regard to machinery in general, in regard to the mass production system we have built up in Germany and America in particular, and people are feeling somewhat bewildered and doubtful as to their value and asking themselves whether we have not, after all, overdone it".[28]

One foreign economist admitted in a talk with the present author some years ago that the people in the West have overdone it.

Why do they think that they have overdone it? What does that expression mean? Does it not imply that after a certain stage machine system begins to bear its evil and destructive fruits? Yes, it does mean that. Gandhiji was far-seeing; he could see the inevitable result of our depending on a system of production that involved the use of power—power that exploits, power that destroys the finer instincts of man. What is true of Germans and Americans is true of all the people of industrially advanced countries of the world. Russia, France, United Kingdom, Sweden Japan and not China are realizing that wealth helps us to satisfy our physical wants but in the process of satisfying them it gives rise to mental and spiritual wants. We have already observed that one cannot walk in opposite directions at the same time. Gandhiji warned us against taking a single step in the wrong direction. Every step that we take in the direction of satisfying our physical wants creates further and further physical wants. We then blindly walk in the same direction, having been intoxicated by the fleeting pleasures of a physically comfortable life. Like men under the influence of liquor we then commit sins, exploit our friends and or brethren, scarcely realizing that we are sowing the seeds of our own destruction. While taking to foreign friends Gandhiji often emphasized the fact that industrialization leads to unemployment. If labourers lose their jobs that is a serious type of exploitation of the poor. Leave aside other evils of the use of machinery; how are we going to solve the problem of unemployment in their both humanitarian and economic aspects? Gandhiji said, "You may say that workers thrown out of work by the introduction of improved machinery will find occupation in other jobs. But in an organized country, where there are only fixed and limited avenues of employment, where the worker has become highly skilled in the use of one particular kind of machinery, you know from your own experience that this is hardly possible".[29]

Machinery can do one or two things; it can either give us more wealth for the same amount of work or give us the same wealth for a smaller amount of work. Actually it tends to give us in some measure both these things. If it gives us more wealth would welcome it, at least from the short-period point of view. In the final analysis, even more wealth is undesirable but from the short-period point of view it might appear to be a blessing. With this reservation the use of machinery, in so far as it gives us more wealth, we might welcome. But then the question of distribution is there. It gives us more wealth but it places this increased wealth in the hands of a few people. This is almost inevitable in the case of a mechanized system of production. Wealth does not pass on from the hand to the mouth directly. There are so many hands and so many mouths in between. Wealth must in such a system flow unevenly and get accumulated and concentrated at certain points. If machines can give us more wealth but distribute it unequally, whatever little advantage there might be in our being able to produce more is lost.

Then let us consider the other thing that machinery can do. It can give us not more wealth but more leisure. Leisure has its own use, if only we know what is good and what is bad use? If, however, leisure is maldistributed as wealth is, if some get very little leisure and others get much more; and if wealth gets distributed on the basis of work, the greater the leisure the lesser is the wealth that one gets. From the macro point of view, it is an advantage to us if machinery can give us more leisure. But its proper distribution is most necessary in a system in which the right to income is dependent on the work we do. In our economic order, therefore, machinery can do nothing that we could welcome.

The main difference between a wise man and others is that the former sacrifices the present for the future while the latter sacrifice the future for the present. A wise man knows that his sacrifice of the present is like an investment which pays handsome dividends in the future. He sacrifices present enjoyment in order to make an investment that is calculated to yield a long-lasting psychic income in the future. An ordinary man, on the other hand, who seeks present enjoyment, unmindful of what may happen in the future, eats up his capital to talk in the language of economics) as it were. That results in the reduction of psychic income in the future. A wise man utilizes the present to undertake an act of production, as it were. That results in the reduction of psychic income in the future. A wise man utilizes the present to undertake an act of production, as it were. An ordinary man treats the present as an act of consumption. He seeks immediate enjoyment while a wise man seeks enjoyment in the future.

A difference that is almost inseparable linked to that already explained is that a wise man seeks methods of *self*-realization. In paying attention to the future he is forced to take note of the fact that lasting enjoyment, which in the final analysis consists of psychic income or spiritual income, depends on taking himself nearer the goal of *self*-realization.

People of the West are now beginning to realize that they had sacrificed the future for the sake of the present. The greatest sacrifice has been that of mental peace. The body has been provided with comforts and luxuries while the mind has been starved of its food — mental peace. Mental peace is enjoyed when the mind diverts its attention from matter to spirit. The ways of living of the people of Western countries have been responsible for encouraging the mind to attach itself to matter. Gandhiji and other men of his stature have succeeded to some extent in making the so-called developed nations realize that we can have comfortable and luxurious living only at the cost of spiritual degradation—degradation that robs the mind of peace and calm.

Now to go back to our main point, Gandhiji said, "A humanitarian industrial policy for India means to me a glorified revival of hand-spinning, for through it alone can pauperism, which is blighting the lives of millions of human beings in their own cottages in the land, be immediately removed....... I would, therefore, have all young men with a scientific training to utilize their skill in making the spinning wheel, if it is possible, a more efficient instrument of production in India's cottages. I am not opposed to the progress of science as such. On the contrary the scientific spirit of the west commands my admiration".[30] (*Young India*, 17th December, 1925).

That was long before Gandhiji left this world. Much water has flown under the Ganges bridge since then. Gandhij never believed in sticking to his views when once expressed. As he said, he was always experimenting with Truth. And he had not found the Truth in 1925; nor did he find the whole of it in 1948. Yet, the more he approached Truth the more he felt that machinery even when it made production more efficient stood in the way of our spiritual progress. Perhaps one who has already realized the *self* can use machines without getting contaminated by it; but then such a person would not need to use it. The less of tools and machines we have the easier it becomes for us to attain our final objective of life.

God has given us a machine; that is our body. It is that machine and only that machine that we must use and by making a proper use of it we can attain *moksh*. God wants us to use this machine that he has given us; the use of other man-made machines interferes with the proper use of the God-given machine.

We have said earlier that the use of machines gives a stimulus to our greed. Read here what Gandhiji wrote in *Harijan* of 26th October, 1934 : "Human greed, which takes no heed of the health or the wealth of the people who come under its heels, is responsible for the hideous rice-mills one sees in all the rice producing tracts. If public opinion was strong, it will make rice-mills an impossibility by simply insisting on unpolished rice and appealing to the owners of rice mills to stop a traffic that undermines the health of a whole nation and robs the poor of an honest means of livelihood".[31]

Gandhiji wrote further in *Harijan* of 10th March, 1946, "The planting such machinery and engines on a large scale in village is also a sign of greed. Is it proper to fill one's pockets in this manner at the expense of the poor? In the thousands of villages where power machinery exists, one misses the sweet music in the early morning of the grinders at work".[32] This amply bears out the truth of what we have said; machinery, besides doing other things that Gandhiji did not like, increases human greed. Man is greedy by nature and this inherent greed is stimulated by the possibility of exploiting other people for one's own benefit. If this is well understood, it will become clear that in no circumstances can machinery be used with impunity. Let us repeat: machines and God are poles apart. Approach machines and draw yourself farther away from God.

Why is exploitation of other people bad? Is it not true that one must look to his own interest first and then, if possible, to the interest of other people? We have already answered this question in the course of our lectures. We have said that in the final analysis no distinction can be made between individual and social welfare. What is good for others is good for you too and *vice versa.*

Now hear the same thing directly from the lips of Gandhiji. "Therefore one who serves his neighbour, serves all the world. As a matter of fact there is in *swadeshi* no room for distinction between one's own and other people. To serve one's neighbour is to serve the world. Indeed it is the only way open to us of serving the world. One to whom the whole world is as his family should have the power of serving the universe without moving from his place……. Tolstoy goes further and says that at present we are riding on other people's backs; it is enough only if we get down. This is another way of putting the same thing. No one can serve others without serving himself. And whoever tries to achieve his private ends without serving other's harms himself as well as the world at large".[33] This is what Gandhiji said in his *Ashram*.

This is what Gandhiji said in connection with his advocacy of *Swadeshism*. *Swadeshi*, therefore, does not involve any disservice to the foreigners. Enlightened selfishness does good to him who performs such an act of selfishness as also to all those who in any way are in association with him. An extension of this idea would lead one to appreciate Gandhiji's advocacy of self-sufficiency. Did he not say that he would like every village and even every individual to be self-sufficient? Breaking off our economic ties with others is tantamount, in the final end, to the strengthening of spiritual bonds along them. If all of us could live together like a big family, of all the individuals of the world could look upon themselves as members of a big family, all of us could live together like a big family if all the individuals of the world could look upon themselves as members of a big family, all would be well. But in the very nature of things such an outlook on the part of people is beyond our imagination. In a joint family of the type that we have in India there are frequent quarrels among the members, similarly there would be quarrels among the peoples of the world on a stupendous scale. And are we not witnessing such quarrels ? If one could concentrate on his own interest – one's own *real* interest – one would be guarding the interests of others. Let us recall what we have had an occasion to refer to elsewhere, namely, that once we reduce ourselves to zero on the material plane, we inflate ourselves to infinity on the spiritual plane.

I am tempted to quote poet Shelley in this connection. He says, "The fountains mingle with the river / And the rivers with the ocean The winds of heaven mix forever, With a sweet emotion; Nothing in the world is single; All things by a law divine in one spirit meet and mingle".[34] (*Love's Philosophy*)

True it is; fountains, rivers and ocean are the same. All things meet in one spirit. Does not the poet say the same thing? Our indwelling spirit, our souls are the same. What appear different are only superficially different; they are different on the surface. How can then one do good to one's self without doing good to others?

And what has Wordsworth to say?

Dust as we are, the immortal spirit grows
Like harmony in music; there is a dark
Inscrutable workmanship that reconciles
Discordant elements.[35] (*The Prelude* I)

True it is; the immortal spirit grows. The external forms of men seem to possess discordant elements. But the inscrutable workmanship of *spirit* reconciles those elements. So men are at bottom the same. Serve yourself and you serve others. And this reminds us once again of what Maharishi Raman said in reply to a question by a visitor. The question was, in what way do you serve others by remaining here and thinking of God? Are you not merely serving yourself? Maharishi put a counter question, "Tell me, what is the difference between me and others".[36]

I have quoted frequently from the writings of foreign poets with a purpose. If I quote always from religious books some of you who do not have an abiding faith in religion and God might not be convinced of the truth of Gandhiji's views. I have, therefore, chosen to quote Western poets who were born and brought up in a civilization that is, at any rate superficially, surcharged with materialism. Even these men when they think deeply about what they see around them come to have a glimpse of the ultimate Truth. Truth, we have already said, is immutable; it cannot be broken into bits, try howsoever you like, you cannot break it into fragments. It is shy as a coy maiden; it hides its head behind a veil but will let you have a look at her when you choose to lift the veil.

Let Gandhiji speak again. "The vow of *swadeshi* is a necessary vow....I suggest to you we are departing from one of the sacred laws of our being when we leave our neighbour and go out somewhere else in order to satisfy our wants. If a man comes from Bombay here and offers you wares, you are not justified in supporting the Bombay merchant or trader so long as you have got a merchant at your very door...... That is *swadeshi*. So when we find that there are many things that we cannot get in India, we must try to do without them. We may have to do without many things which we may consider necessary; but believe me, when you have that frame of mind, you will find a great burden taken off your shoulders, even as the Pilgrim did in that inimitable book, *Pilgrim's Progress*. There came a time when the mighty burden that the Pilgrim was carrying on his shoulders unconsciously dropped from him, and he felt a freer man than he was when he started on the journey. So will you feel freer man than you are now immediately you adopt this *swadeshi* life"[37]. (*Speeches and Writing of Mahatma Gandhi*, Natesan, 4th Ed.)

This is the same burden of the song: be self-supporting and if you cannot, depend upon those who are close to you. Does this not elaborate the same point, namely, that by serving your own interest you serve others too? If every man could do that nobody would remain unserved. Gandhiji did not advise only his country men to take the vow of *swadeshi*; he wanted everybody, wherever he or she may be, to adopt the *swadeshi* life. Charity must begin at home. And if it begins at home it does not end there. Charity in every home would total up to charity in the whole world.

"Under this plan of life", said Gandhiji, "in seeming to serve India to the exclusion of every other country, I do not harm any other country. My patriotism is both exclusive and inclusive. It is exclusive in the sense that in all humility I confine my attention to the land of my birth, but is inclusive in the sense that my service is not of a competitive or antagonistic nature."

No certainly it is not competitive or antagonistic. We have said again and again that a man's first duty is to serve himself. If he cannot serve himself he cannot serve others. His next duty is to serve those who are near and dear to him. If he does not do that, well, there is none whom he really loves. In these few words that Gandhiji wrote he has expressed in simple words Truth of the highest order. "Verily I say unto you, Inasmuch as ye have done it unto one of the least of these my brethren, ye have done it unto me".[38] (New Testament, *Mathew*, XXV)

Gandhiji's *swadeshism*, as he said, consists in helping and loving one's neighbour. We have seen how when one helps one's neighbour the whole world comes to get the needed help. Thou shalt love thy neighbour as thyself (*St. Mathew*). If only everybody loved his neighbour the whole world would be engulfed in love. So in the holy Bible we find the same truth that Gandhiji found by his experiments.

In the end let us see what Gandhiji has said against modern machine-civilization and how he has eulogized the ancient Indian way of living. "Our ancestors, therefore, set a limit to our indulgences. They saw that happiness was largely a mental condition. A man is not necessarily happy because he is rich, or unhappy because he is poor......... Observing all this our ancestors dissuaded us from luxuries and pleasures. We have managed with the same kind of plough as existed thousands of years ago. We have retained the same kind of cottages that we had in former times, and our indigenous education remains the same as before. We have had no system of life-corroding competition. Each followed his own occupation or trade and charged a wage. It is not that we did not know how to invent machinery, but our forefathers knew that if we set our hearts after such things, we would become slaves and lose our moral fibre. They, therefore, after due deliberation decided that we should only do what we could with our hands and feet".[39]

So in very clear words Gandhiji has expressed himself against the use of machinery, giving us the example of what our forefathers thought and did. It brings out the truth of what we have said, namely, that the use of material aids to production places a hurdle in the way of our efforts to realize our *self.*

"Our forefathers saw that kings and their swords were inferior to the sword of ethics and they, therefore, held the sovereigns of the earth to be inferior to the *rishis* and the *fakirs*.[40] (We have quoted this from *Hind Swaraj*, 1962 ed.)

I take the liberty of quoting from *The Philosophy of Life* by Anderson M. Baten where he uses the words of John Temple Graves: "No blaze in all our eulogy can burn besides the sunlight of a useful life. After all there is nothing grander than noble living."[41]

"I have seen the rays that gleam from the headlight of some giant engine rushing onward through the darkness of opposition, fearless of danger, and I thought it was grand.

"I have seen the light come over the eastern hills in glory, driving the hazy darkness like mist before the sea-borne gale, till leaf and tree and blade of grass glitter in the myriad diamonds of the morning ray, and I thought it grand.

"I have seen the light that leaped at midnight, athwart the storm-swept sky shivering over chaotic clouds, mid howling winds, till cloud and darkness and the shadow haunted earth flashed into mid-day splendour, and I knew it was grand.

"But the greatest thing next to the radiance that flows from the almighty throne is the light of a noble and beautiful life wrapping itself in benediction around the destinies of men and finding its home in the blessed bosom of the everlasting God."[42]

True it is; nothing is so grand as the light of a noble and beautiful life that finds its home in the blessed bosom of the everlasting God. The life that is lived by people who have enslaved the force of Nature and who with their help seek to enslave their brethren can never find its home in the bosom of the everlasting God. The roads are different but they all lead to the same destination: great men are alike in one grand respect that in their moods of deepest meditation they come to think alike.

And to wind up, here is what Elbert Hubbard has said (quoted from *Felix Shay's Book*) "Men are homesick amid this mad rush for wealth and place and power. The calm of the country invites and we would fain do with less things and go back to simplicity.[43]

Before we finally wind up our study of Gandhian thought on Machinery, let us see what people who do not think in terms of realization of *self* have had to say about machinery. Human history abounds in instance of man's fear of his power of invention and the consequences of his defying and enslaving the forces of nature. Today with the greatest victory that man has gained over dumb Nature, he has begun awfully to realize that his intelligence, devoid of wisdom, is most likely to destroy the entire human race. This may not actually happen; who knows what the infinite power has ordained for man. But the fear is there and is a genuine one too.

In the present age characterized by the supremacy of technology over other more delicate human instincts, it is thought that science is going to be our dictator. To quote from the book *Ideas* published by the Waverley Book Company Ltd., London, "While relieving mankind of a great deal of sheer drudgery, the machine has imposed upon its great deal of sheer drudgery, the machine has imposed upon its creator a new form of drudgery, namely the increasingly exacting task of maintenance. The leisure which the machine-age was supposed to afford mankind has been mysterially whittled away. Instead of enjoying more time for social intercourse with his fellows, man spends a great part of his leisure in listening to and looking at machines.

"Even the time that he saves is unproductive; we move heaven and earth, it has been said, to save five minutes and then have nothing to do with them when we have saved them. All this points to an inescapable conclusion: machines have imposed a new *rhythm*, if not a new tyranny, upon our lives. This is the reason why advocates of the Simple Life almost always recommend as a first step to freedom the abolition of machinery.

"*Assuming that the development of Cybernetics produces apparatus of even more remarkable range and scope, man will still be master of his creations. But he will be a master increasingly burdened with technical responsibilities. It will be true of him, as of no tyrant of antiquity, that the master is the slave of the slave. He will have to make up his mind whether a mechanistic civilization makes life intolerable, or whether it leaves room for.*

Religious men, poets, philosophers and scientists have all realized that our machine-civilization is bound to end in the annihilation of the human race or at any rate it is bound to make our life intolerable. And yet man seems to be helpless in doing what he thinks he should; there is a destiny that mars our ends.

"*Even Gandhi, with all his charisma, did not melt the hearts of his oppressors, as he had hoped. After softening, hearts harden again. Asoka too was wrong to think that he was changing the course of history, and that his righteousness would last 'as long as the sun and the moon'*." (Theodore Zeldin in *An Intimate History of Humanity*, 1994)

Gandhi has been criticized for his nostalgia for ancient rural bliss and delaying the modernization and industrialization of his country. On the other hand, he has been regarded as the "true soul" of India. With other Hindu sages Gandhi shared a mistrust of worshipping followers, and he tried to avoid the title *Mahatma*. In spite of this, his disciples regarded him as a saint. Gandhi's denial of the pleasures of food, sex, family, and friendship, has made his way of life extremely demanding for ordinary people, who otherwise have found inspiration from his courage and teachings. This question troubled *George Orwell* in his essay *Reflections on Gandhi* (1949). While admitting that Gandhi never made claims of sainthood, he did not hesitate to reject sainthood as an ideal: "The essence of being human is that one does not seek perfection, that one is sometimes willing to commit sins for the sake of loyalty, that one does not push asceticism to the point where it makes friendly intercourse impossible, and that one is prepared in the end to be defeated and broken up by life, which is the inevitable price of fastening one's love upon other human individuals".

"*Become the change you want to see in the world.*" (Vice President Al Gore's quotation from Gandhi)

For further reading: *Gandhi's Passion: The Life and Legacy of Mahatma Gandhi* by Stanley Wolpert (2001); *Gandhian Satyagraha: An Analytical and Critical Approach* by Ajay Shanker Rai (2000); *Gandhi's Body: Sex, Diet, and the Politics of Nationalism* by Joseph S. Alter (2000); *Colonialism, Tradition, and Reform* by Bhikhu C. Parekh (1999); *Gandhi in the Twenty First Century: An Alternative Approach to Development*, ed. by Vasant Kumar Bawa (1999); *The Forgotten Woman: The Untold Story of Kasturba Gandhi, Wife of Mahatma Gandhi* by Arun Gandhi, *et al.* (1998); *Gandhi: A Life* by Yogesh Chadha (1998); *Gandhi Great Soul* by John B. Severance (1997); *Gandhi: Struggling for Autonomy* by Ronald J. Terchek (1998); *Gandhi: The Power of Pacifism* by Catherine Clemen (1996); *The Gandhi Reader: A Sourcebook of His Life and Writings*, ed. by Homer A. Jack (1995); *Gandhi: Prisoner of Hope* by Judith Brown (1990); *Mahatma Gandhi: A Biography* by B.R.

Nanda (1989); *The Moral and Political Thought of Mahatma Gandhi* by R.N. Iyer (1973); *Conquest of Violence* by Joan V. Bondurant (1965); *Mahatma* by Dinanath G. Tendulkar (1960-63, 8 vols.); *Mahatma Gandhi* by Bal Ram Nanda (1958)

For further information: *Epigrams from Gandhiji—Man of the Millenium—Profile—Mahatma Gandhi* (1869-1948).

Appendix 2

Gandhi: Sustainability and Technology as Ideas of Political Practice

*Anup Sam Ninan**

The author here attempts to draw the attention on how an intense political movement tried to address the question of technology while defining its praxis. It is particularly of relevance at least in the following grounds; (i) as how mass movements negotiate and appropriate different definitions and attributes of technology, (ii) as how a movement nearly a century back looked at certain political issues associated with technology that are still considered to be core issues, and (iii) as to look at how these articulations were located in the discursive terrain of colonial experience.

The proposals are temporally located at the colonial and late-colonial period of Indian history, more specifically in the first-half of the twentieth century. Though most of the studies consider Mohandas Gandhi as an exclusive figure while discussing what is called as Gandhian School of Thought, this paper incorporates the views of J.C. Kumarappa, one of the most vocal advocates of Gandhian economic and technological programme, and Vinoba Bhave, the 'spiritual successor' of Mahatma Gandhi, also to observe the collective assumptions of the school while it also posit the centrality of Mohandas Gandhi.

It forms a part of the wider articulations during the period about the possible visions of future independent India. While exploring the underlying notions of progress, selfgovernment and national building that contained in the development goals that articulated in the late-colonial period, Benjamin Zachariah (2005) identifies three distinctive groups who were part of the process. There were the British Indian administrators, the Gandhians and the other nationalists referred to as modernizers.

Here it is emphasized one of the groups for its exploration and attempts to observe how its protagonists constructed the narratives of science and technology as part of their political practice. It does not intend to assess how this rhetoric was translated into the practical schemes by the promoters or how it was contested by the contending ideological categories. Despite the fact that the Gandhian proposals of development were met with considerable challenge by the 1930s, and that it almost given its way to the 'modernizing' school by 1950s, Here we try to emphasise a part of the intense and instructive phase of a debate wherein the fundamental aspects of science and technology were under close scrutiny.

Mohandas Karamchand Gandhi who was born in Gujarat in 1869 and was assassinated at Delhi in 1948 was an important figure in the early twentieth century politics of Indian subcontinent. Born in a wealthy upper-caste family, he was trained as a lawyer in Britain and practiced as an attorney in South Africa. While in South Africa, a personal experience of apartheid, as being thrown out of a first class train compartment despite having valid ticket, triggered his entry to active politics. His basic ideas of political practice emerged from the experience he accrued by organizing the Indians in South Africa. In 1914 he returned to India and by 1919 he emerged to be one of the significant influences in the nationalist movement against British colonialism. He spearheaded the transformation of Indian National Congress, till then a platform of elites and educated Indians resorting to formal and constitutional means of freedom struggle, into a mass movement. He was also instrumental in bringing about the freedom movement as a nationalist movement comprised of rural peasantry and the masses from being an elite bourgeois activity (Embree 1989:165; Khilnani 1998:125)

The Context

The colonial experience that got rooted from the seventeenth century had drastic implications for the economy and society of the subcontinent. While the values of enlightenment and modernity that was part of the colonial project brought significant alterations to the society and polity, the economic impact of colonialism was deeper. It transformed a region that was among the most prosperous in the world till fifteenth century to one of the poorest regions in the passing of two to three centuries.

Though there were sporadic and disconnected resistances against the colonial powers from the beginning, the organized attempts against the colonial rule emerged only in the latter half of nineteenth century. By then, there emerged a elite of which one significant section demanded for the end of colonial rule. They resorted to mobilize the people by instilling the values of confidence by glorifying the 'golden' past and 'civilizational' strengths of Indians. On the other hand, in a simultaneous process, the colonizers were vilified and ascribed them the root cause for all prevailing ills. All the different sections, including the supporters of colonial social reform projects, nurtured the post-colonial aspirations of developed India (though their definitions of development varied drastically). It is in this larger discursive frame Mohandas Gandhi emerges to political mobilization.

Unlike his predecessors and contemporaries, Gandhi attempted to address wide ranging readers/audience/activists. He made village as an important ideological category and the focal centre of struggle. Since the prevailing national movement was predominantly urban and elitist, Gandhi set to cater entirely wide and different categories of population that pose a methodological problem in assessing his works. His usage of terminology and ideas contextually varied to an extent that it may overtly looked contradictory sometimes. However, it is to be noted that, he employs the rhetoric of science and technology wide across the different sections, as a major issue in political debate.

Another problem we face while analyzing the works of Mohandas Gandhi is the centrality of 'Mahatma' (the great soul) Gandhi. The centrality of the personality poses two particular hurdles. The individualized ideology incited the followers to dwell into the problems and solutions Gandhi considered to be important and even his prejudices were spilled over to the majority (caste, gender, vegetarianism). It also prompted most of followers, later day analysts and commentators to confine on his works alone while discussing about it. Similarly, the terminology Gandhi uses blurs the concepts mutually. It overlaps across ideas of science, technology, machine, machinery, modern civilization, etc. Moreover, his views are also changed over a period of time. On the same vain, spirituality and blatant superstitions are also found across his writings. It falls somewhere between his personal beliefs and conviction of the specific audience he addresses (being a very shrewd politician). Given the above background and disclaimers, let us look at how the issues of our concern were dealt.

Technology as a Problematique in Politics

The Gandhian school of thought stood for a decentralized, self-reliant village based economic order with labour intensive rural and small scale agriculture and industries with the support of low technologies wherever required. The book, *Hind Swaraj* (Indian Home Rule) published in 1909 (and banned in India in1910 by the British Government for sedition) forms the basic document of Gandhian thought. Gandhi poses a staunch attack on the western development model that he calls as the culture of machines.

Machinery has begun to desolate Europe. Ruination is knocking at the English gates.Machinery is the chief symbol of modern civilization; it represents a great sin (Gandhi 1909)

Machinery is like a snake-hole which may contain from one to a hundred snakes. Where there is machinery there are large cities; and where there are large cities, there are tram cars and railways; and there only does one see electric light. English villages do not boast any of these things......I cannot recall a single good point in connection with machinery. Books can be written to demonstrate its evils. (Gandhi 1909)

However, as he becomes an active participant in Indian politics, by 1920s the Luddite view is under change, and that remained more or less consistent throughout the rest of his life.

"*I have nothing to withdraw from what I said about machines in the Indian Home Rule......I do not suggest today a destruction of all the machines, but I am making spinning wheel the master machine. The Indian Home Rule depicts an ideal state.*" (Gandhi, 1924)

By mid-1920s Gandhi's views get more balanced and reasoned:

"*I had no quarrel with rail roads, steamers and many machines as such, but that I protested against the abuse that was at present being made of them, either for exploiting many nations of the earth or for destroying them.*" (Gandhi, 1925a)

It gets further progressed to accommodating other issues particularly the labour displacement.

"*Machinery has its place, it has come to stay. But it must not be allowed to displace the necessary human labour……I would welcome every improvement in the cottage machine, but I know that it is criminal to displace the hand labour by the introduction of power driven spindles unless one is, at the same time, ready to give millions of farmers some other occupation in their homes.*' (Gandhi, 1925b)

When situations demand, he does not even averse of importing technology, as in the following case:

"*As long as we cannot make the machine required for utilizing the hide of dead cattle, worth nine crores, available in our country, I would be ready to import them from any part of the world.*" (Gandhi, 1927)

As his earlier views about machinery attracted critical attention in almost all occasions, he used to provide specific messages on those grounds, and often reminded that what he resisted is the usurpation by machinery of the functions of humans and their consequent slavery to it (Gandhi, 1929)

Similarly, in an interview to Edmond Demeter in 1931, he clarifies, "*I am an enemy not of mechanism but of organized mechanism. I consider this system, which has become the basis of your civilization, as the greatest danger which could menace man*". (Gandhi, 1931)

He further attempted to contextualize his viewpoints with more systematic reasoning:

"*Mechanization is good when hands are too few for the work intended to be accomplished……It is an evil when there are more hands than required for the work, as is the case in India.*" (Gandhi, 1934)

As he is more and more deep into the political terrain, and due to the partial waning of his ideas among the nationalists, he becomes more encompassing in his idea about technology as time passes by:

"*I do visualize electricity, ship-building, iron works, machine making and the like existing side by side with village handicrafts. But the order of dependence will be reversed. Hitherto industrialization has been so planned as to destroy the villages and the village crafts. In the state of future it will subserve the villagers and their crafts.*" (Gandhi, 1940)

Even when mapping the transformation in his viewpoints one observe the significant space Gandhi attributed to technology in his political practice. Technology was a crucial factor/agent in articulating issues ranging from democracy, representation, progress, employment and equity. He considered choice of technology as the most crucial element in determining the production and distribution processes. The school, through its different organizations like All India Spinners Association and All India Village Industries Association, both formed in the 1930s, proposed decentralized small scale industrial models that encapsulated the ideals like direct democracy, localized production and distribution, self-reliance of the community and thereby the nation, and political and technological emancipation through localized actions. At the same time, it began to recognize the inevitability of large-scale industries also.

"*Large scale industries are needed in exploiting natural resources which should form the raw materials of the people's industries. Control of such raw materials cannot be left in private hands. Therefore, all such industries should also be under the state. In conomics, large scale industry is the antithesis of democracy in politics. It is not by chance that the western nations have come by their economic organization. It is a result of their way of thinking in terms of autocracy. They find themselves with dictatorships in political organization, and centralized industries in the economic field. These two go together and we cannot have the one without the other. If we aim at democracy, we have to abandon centralized production in consumption goods.*

"*This does not preclude all centralized industries. As we have already seen there is a definite field of such, though restricted.*" (Kumarappa, 1945)

Technological Metaphors in Political Practice

Technological artifacts and metaphors have played a significant role in Gandhian political practice. It symbolized political meanings; it was through artifacts like spinning wheel and handloom cloth the deeper political messages like self reliance and independence were conveyed to the masses. It was symbols, processes and products of grassroots mass mobilization.

"*The movement of spinning wheel is an organized attempt to displace machinery from that state of exclusiveness and exploitation and to place it in its proper state. Under my scheme, therefore, men in charge of machinery will think not of themselves or even of the nations to which they belong but of the whole human race.*" (Gandhi, 1935)

"*You cannot build non-violence on a factory civilization, but it can be built on self-contained villages. Even if Hitler so minded, he could not devastate seven hundred thousand non-violent villages. He would himself become non-violent in the process. Rural economy as I have conceived eschews exploitation altogether and exploitation is the essence of violence. You have, therefore, to be rural minded before you can be non-violent, and to be rural minded, you have to have faith in the spinning wheel.*" (Gandhi, 1936)

The school, thus, appropriated science and technology as a means of routine political activity. It integrated the technological process to the larger political process while infusing political ideals to artifacts and its design.

Contextual Relevance (Problematising Technology)

Gandhian school gave primacy to the process, users and control of technology. Centrality and proximity of makers and users is of primary importance in this conception. Labour-intensiveness is considered to be a basic feature in the notion of technology and widely across, it is argued that technology should not replace workers in a populated country.

"*There is nothing good or evil in inanimate things. These attributes come into existence by man's reaction to these. A thing is good or evil according to as we put to benefit or harmful use...... Machines can be masters as well as servants......Machine or no machine is not the fundamental question. The problem is one of finding the best way of not only supplying the material wants but also bringing employment to the people. Machinery which only saves labour is a curse in an unemployed or underemployed India, whatever it may be in other countries. Machinery can only be used to make such things as hands cannot make and where will not displace workers gainfully employed.*" (Kumarappa, 1936:99)

Again, knowledge and technology are conceived to be common property and in case of large industries, it was argued that it should be state owned.

"*In the village movement there is a definite place for centralized industries, not for their own sakes, but as adjuncts and subsidiaries to decentralized units, just as in the political sphere a democracy does not eschew governmental control and regulations when it is directed towards the better realization of individual development and expression.*" (Kumarappa, 1936:101)

Besides that, the school vociferously argued and devised methods to emphasise that science and technology practices impacted day-to-day lives of the ordinary citizens. Thus, it devised 'scientific' methods not only to spinning, but also to sanitation, diet improvements, etc.

Human, Animal and Nature Relations

Quite similar to the above lines, human beings and animals are believed to have organically linked to the nature. It is argued that the exploitation inflicted on animals and nature in the name of progress is antithetic to the healthy co-existence. For instance, Gandhi considered scientific research based on vivisection as subjugation of non-human world.

I abhor vivisection with my whole soul. I detest the unpardonable slaughter of innocent life in the name of science and humanity so-called, and all the scientist's discoveries stained with innocent blood I count as of no consequence (Gandhi, 1928)

Similarly, it attempted 'scientific' improvisation to the traditional medicinal systems of Ayurveda and Unani.

Observations

Gandhi attempted to give alternative reading to science and technology in the colonial context. For that he questioned the method of scientific and technological practices, objectives (particularly posing moral arguments to it) and also trained a new range of practitioners who were called as scientists. However, he resorted to the same tools and yardsticks of modernity to question the features of colonial modernity, though it is widely held otherwise.

Secondly, with the emergence of post-colonial studies, Gandhi is depicted to be a cultural icon who posed political epistemological challenge to western hegemony. However, a closer observation of Gandhi reveals that he was more in tune with the criticism of industrial capitalism and colonialism emerged wide across, including in the west. Tolstoy, Thoreau, Ruskin, and Edward Carpenter influenced his thoughts at a significant level.

REFERENCES

Anup Sam Ninan; *Gandhi—Sustainability and Technology as Ideas of Political Practice.*

Embree, Ainslie Thomas (1989). *Imagining India: Essays on Indian History,* New Delhi; Oxford University Press.

Gandhi, Mohandas Karamchand (1931). Interview to Edmond Demeter, *Collected Works of Mahatma Gandhi* ,Vol. 48, p. 386.

Gandhi, Mohandas Karamchand (1909). *Hind Swaraj or The Indian Home Rule* 1938 Ed., Ahmedabad; Navajivan Publishing House.

Gandhi, Mohandas Karamchand (1924). Navajivan 10-8-1924, *Collected Works of Mahatma Gandhi,* Vol. 28, p. 548.

Gandhi, Mohandas Karamchand (1925a). *Young India*, 18-8-1925. *Collected Works of Mahatma Gandhi,* Vol. 28, p. 63.

Gandhi, Mohandas Karamchand (1925b). *Young India*, 5-11-1925. *Collected Works of Mahatma Gandhi* Vol. 28 p. 428.

Gandhi, Mohandas Karamchand (1927). *Navajivan* 19-6-1927. *Collected Works of Mahatma Gandhi.*

Gandhi, Mohandas Karamchand (1928). *Collected Works of Mahatma Gandhi* 29; 325.

Gandhi, Mohandas Karamchand (1929). *Young India,* 21-3-1929. *Collected Works of Mahatma Gandhi* Vol. 40 p. 165.

Gandhi, Mohandas Karamchand (1934). *Harijan,* 16-11-1934.

Gandhi, Mohandas Karamchand (1935). *Young India,* p. 321.

Gandhi, Mohandas Karamchand (1936). *Collected Works of Mahatma Gandhi,* 70; 295.

Gandhi, Mohandas Karamchand (1940). *Harijan,* 27-1-1940.

Khilnani, Sunil (1998). *The Idea of India*, New York; Farrar Strauss Giroux, 6th Annual IAS-STS Conference: Critical Issues in Science and Technology Studies, 24-25 May 2007, Graz.

Appendix 3

The Problems of Today as seen by Mahatma Gandhi and Solutions by Param Pujya Gurudev

This blog is based on a talk I delivered on "Technology—The Passion and Responsibility"at the National level symposium on "The message of Hind Swaraj and its contemporary relevance" at Madurai on 27th February 2010. The talk was based on a book *Hind Swaraj* written by Mahatma Gandhi in 1908. I spoke on a chapter about machinery. This wonderful book is attached here.

Today, technological innovation is considered a modern day marvel which has improved the living conditions of people. It is virtually impossible for people to imagine a life without using technology.

Technology is essential to physical existence of humans. Fire, clothing, farming is all part of technology. It is technology that enabled the hasty evolution of life on earth.

There are two types of machine used :

1. Those that aid human efforts; and
2. Those that reduce human efforts.

Machines like telescopes, microscopes allow you to see far and smaller objects. The potter's wheel allows you make pots that are stronger and uniform. The telephone allows you to talk to people far away. The car lets you travel fast. The grinder lets you grind fine. Machines like a crane or JCB lets you lift heavy weights. The pressure cooker cooks your food fast. The microphone lets our voices go across the room. Today technology allows you to fly in the air, go into space as a tourist, go to the depths of the ocean, technology allows you to do your entire genome decoding, create satellites around the earth, build a space station, technology allows you to place stents in arteries, destroy cancer cells with chemotherapy, you use computers in a multitude ways to communicate, compute, record, play, document, entertainment, etc.

Certainly, life has changed in the past 50-100 years in such a manner as it has done never before. The main reason for this exponential growth is the advancement of technology.

Technology allows us to develop better technology. This is a self-serving cycle that is the cause of all these wonders. Technology makes life easier

Think of how would you meet a stranger at a park a few years earlier. You would plan the visit, identify the place where you would meet, identify the exact location, the exact time, the modes of identifying each other and then go and wait for the other person.

Today, you set a time and place and then keep calling each other over phone on where one is. We probably have forgotten the art of planning.

Technology brings awareness. One of the advantages of modern technology is that it brings the world to your door – it brings more means of entertainment to us. While this is certainly true, how has this been beneficial for your individual growth?

We are aware of what is happening around the world. This widens our horizons. (We see the same violence in all places, the same natural disasters, at times we see better responses to disasters, but seeing it, are you trying to better our situation? In that case, it will be useful.) For most people, their life will remain unchanged if their televisions stopped working for a couple of months.

Entertainment is something that turns stale very soon. While you might love a movie or a genre of movies, you will not be interested in watching the same thing again and again. It always seeks newer things.

The question that we have to ask ourselves is "Have we really improved our lives with technology?"

After all, modern technology has its basis in the science whose forefather Francis Bacon said "… nature had to be hounded in her wanderings, bound into service and made a slave. She was to be put in bounds and the aim of the scientist was to torture nature's secret from her …". The modern science and technology has evolved with this basis and has succeeded in achieving this aim.

- Technology has brought with it pollution.
- Industries where people work from morning to night.
- The number of jobless people is increasing.
- The economy is growing, but there are more poor.
- We spend more time working than with family.
- Our stress levels have gone up – not reduced.
- We have medicines for diseases and yet there are more diseases that we don't know how to cure.
- Americans have forgotten the art of cooking a meal. Their grand dream was to just pop a pill that has all nutrients and avoid cooking completely. With packaged meals, their health habits are worsening. The social leaders are trying to convince people to cook meals at home with fresh nutrients.
- Our air, water, food is contaminated with no means of cleaning them up.
- We have better weapons that can annihilate the planet.
- Are our children better equipped to handle the world? Why do we have so many suicides?
- You see workers labouring from morning to night earning a wage to sustain themselves.
- Both parents have to compulsorily work nowadays in the cities to make both ends meet. Who then takes care of one of the most sacred duties – to bring up one's children well?
- A ratio of thousand to one for jobs.
- A lack of dignity in traditional jobs – The story of an Andhra engineer. He gave up an IT job to take up farming when his father who was a farmer died. He eventually had to leave farming and go back to IT because he was not getting a bride.
- Growing disparity between rich and poor.
- Loss of jobs because something can be done cheaper by machines.
- The founder of the company I work for said this—"the purpose of business is not just about creating stockholder wealth. It's about meeting the needs of all the constituents—employees, customers and stockholders—in a scenario where everybody wins". How many companies can claim that they work with this model? The primary mode of operation is selfish – to succeed individually.

It is in this context that we have to understand what Gandhiji is talking about. His comments were made when the industrial revolution was just opening in India. His comments created a controversy then and would have created an even bigger controversy today. But his comments were relevant then and relevant even today.

Gandhiji wrote "Machinery has begun to desolate Europe. Ruination is now knocking at the English gates. Machinery is the chief symbol of modern civilization; it represents a great sin. .... The craze is for what they call labour-saving machinery. Men go on "saving labour" till thousands are without work and thrown on the open streets to die of starvation ..."

To understand why this is happening, we should understand who we are and what are we doing on this planet.

Technology and Development

Primary focus of people has now become to work and relax. Personal growth is considered in terms of money and progress in one's career. Progress on the front of personal, psychological, emotional and mental development is not considered at all. We lack both the will of developing and the means of measuring progress on this front.

Many of us define our lives and achievements in our lives by the materialistic gains we have made. We talk about our salaries, the vehicles we own, the house we own, the vacations we take, the multitude of ways we spend money. But do we define the progress in our lives by the actual metrics of progress? Do we look at the increase in empathy, nobility, love in our lives? Do we look at increase in our mental capacity? Do we look at our social responsibility?

When a kid grows up, every year we look at how he/she is developing. Though these days we actually measure the ability of the kid to memorize rather than understand, we can say that there is some form of measurement however imperfect it may be. Unfortunately, this measuring of personality development ceases quickly. This personality development stops in the early twenties for an average person. Today's expected lifespan is 70 years. If we are spending only the first 20-25years of our life and then stop, it is a complete waste of human potential.

When you look at successful politicians, actors, businessmen you see that many of them are active and ambitious even at an age of 70+. This zeal and ambition makes them successful. But an average person's aim would be to get a job, marry, have kids and probably own a house. Is this what is needed in human development?

Scientists tell us that a successful man today uses only about 7 per cent of the total brain's potential. Today, humans are the pinnacle of evolution. We are the best evolved We are all proud of it. But is this the end? Is this what we could achieve?

Scientists think and many scientific spiritualists claim that humanity is just a transitory phase in evolution. They talk of all the potential achievements that can be made.

One example is the advanced senses of *rishis* (Astronomy in Vedas). This has Gandhiji said, "Impoverished India can become free, but it will be hard for any India made rich through immorality to regain its freedom. I fear we shall have to admit that moneyed men support British rule; their interest is bound up with its stability. Money renders a man helpless. The other thing which is equally harmful is sexual vice. Both are poison. A snake-bite is a lesser poison than these two, because the former merely destroys the body but the latter destroy body, mind and soul. We need not, therefore, be pleased with the prospect of the growth of the industry"

Even though Gandhiji rejected the craze for machinery, we all live in a world where machinery have become commonplace. His warnings fell on deaf ears and now we have to salvage this. What is our responsibility now?

Now that we see that we are stuck with the problem, we have to identify how to solve this problem. One of the greates seers of our age, Pandit Shriram Sharma Acharya, has given systematic and simple techniques,i.e., the 5 *sheelas* and 4 *varchas*, which if applied to our daily life, helps humanity come out of this quicksand. We will discuss these techniques during the course of this talk.

Mitavyayata

Today, we are all slaves to machinery. If the machinery around fails to work we feel lost. Check what happens in your life if the machines – geysers, cars, fans, air conditioners, television, phones stop working.

The only way to salvage the situation is to limit their use to only such places where it is absolutely essential. (*mitavyaya* on physical, emotions and thoughts so that you can use *swarajya*) By controlling the use of machinery, we gain control and cease to be slaves of machinery.

Shramasheelata

One of the popular modes of keeping fit is to jog in the morning. Would a person who walks to work and back need this? It is needed to only those who have stopped using their limbs. Today, many follow *yoga asanas* for health. We see that *yoga* teachers teach us *asanas* that were actually part of one's routine if the machines were not there. For example, if you have seen the *asanas* taught by Baba Ramdev, there are exercises like using the grind stone, sweeping the floor, washing clothes, etc. Why are we doing these now? We need them now because these physical activities have stopped in our daily lives. As Gandhiji predicted above, the unrestrained use of machinery has led to atrophy of the limbs and is causing grave health concerns.

The only way to salvage the situation is to start physical activity in our day-to-day activities. We should avoid looking for means to reduce physical activity and instead adopt physical activities for a healthier lifestyle.

What does this mean to our lifestyle? If you are taking the lift to climb two floors, start taking the stairs. Walk to your nearby shop instead of taking the vehicle. If you have no health problems, start taking bath with regular water instead of hot water. You will save energy and also feel more refreshed.

On the other hand, if you have to take a patient to a doctor, do not think that you will make him walk. Take him in the fastest mode possible. Frugality does not mean poverty. Frugality means to use only that which is essential.

When you buy a bigger TV, see why are you buying it. Are you buying it because your neighbour has one or is it because your existing one does not work? Many people buy huge TVs even though their living rooms are not big to watch them well. They end up watching grainier pictures.

Do you really need to take a vehicle to your neighbourhood shop?

What are you gaining by taking a vehicle to work instead of walking?

- You could stay in a calmer part of the city
  - Lets me meditate in peace
- You could stay in a cheaper part of the city
  - Lets me contribute to the society
  - Lets me buy a bigger TV

- I have to take a vehicle because my home is near to a better school for my kid.
- I have no choice – I spend hours travelling everyday because the city is big

Why do you build a factory?

- So that I can build better products
- So that I can reduce the number of people needed to work
- Produce iron so that people can build sturdier houses.
- So that I can save money and make higher profits
  - What do you do with the profits? – Buy more land and build a bigger house…
  - Goes to charity – Tata sons: More than 70 per cent of the shares are held by charities. The profit that these companies make are contributing to the betterment of the society.

Shishtata

Courtesy towards one another. When a person thinks that the entire world is his family – *vasudhaiva kutumbakam*—he learns to love and respect the others. If you love others, you will become less selfish.

Example: Teeth Biting the Tongue

As Gandhiji said, the root of the problem is not machinery but the craze for machinery. "How can I be when I know that even this body is a most delicate piece of machinery? The spinning wheel is a machine; a little toothpick is a machine. What I object to is the craze for machinery, not machinery as such. The craze is for what they call labour-saving machinery. Men go on "saving labour" till thousands are without work and thrown on the open streets to die of starvation. I want to save time and labour not for a fraction of mankind but for all. I want the concentration of wealth, not in the hands of a few, but in the hands of all. Today machinery merely helps a few to ride on the backs of millions. The impetus behind it all is not the philanthropy to save labour, but greed. It is against this constitution of things that

I am fighting with all my might. The supreme consideration is man. The machine should not tend to atrophy the limbs of man. For instance, I would make intelligent exceptions.

Take the case of the Singer's sewing machine. It is one of the few useful things ever invented, and there is a romance about the device itself.

It is the specific types of machinery that trigger laziness that is the problem than the ones that aid human development. All machines fall into this category. It is how you use them that result in the difference.

At a deeper level, remember that the human life is a transitory phase in evolution. It is just a temporary stage between animal and superhuman stages of evolution. The human body is a very powerful and delicate piece of machinery that allows life to evolve. Even this body is to be discarded as one moves ahead in evolution. That is the grandest dream of Gandhiji and many spiritual beliefs.

As long as man is a slave to his body, he is tossed and thrown about as the body wishes. However, when he becomes the master of his self, when he attains "*swaraj*", does he attains *swaraj*. When Indians remember this truism and adopt it, we will have a true *hind swaraj*.

.rajiv's blog

Hmm. I would say don't bother

Hmm. I would say don't bother replying to those posts if they irk you. Check out the person's profile to see how long they have been a member if you're concerned they might be opportunists....... and don't reply to one of those topics *cisco exam questions* if they joined that day or a week before perhaps?

Posted by shazlovely on Mon, 05/24/2010-02:07

Appendix 4

Agenda 21 Section IV and Gandhian Philosophy: A Comparison

Ashok Bhise

Rio Declaration on Environment and Development, Forest Principles, Convention on Biological Diversity, Framework Convention on Climate Change (UNFCCC) and Agenda 21. Agenda 21 is a blueprint for sustainable development into the twenty-first century. There are 40 chapters of Agenda 21, divided into four main sections. It includes combating poverty, changing consumption pattern promoting health, atmospheric protection, combating deforestation, conservation of biological diversity (biodiversity), and control of pollution, the roles of children, youth, women, NGOs, local authorities, technology transfer, Promoting Education, Public Awareness and Training, etc. Agenda 21 has brought out many issues. It is a vast document. However, due to space restrictions only two chapters from Section IV are taken for comparison with Gandhian philosophy.

It is striking to note that most of the issues forming integral part of Agenda 21 are already brought forth by Mahatma Gandhi a century back when there was absolutely no environmental degradation and pollution. All the problems were anticipated long ago and given probable solution to it. There may be repetition of certain points and quotations at more than one place because of their relevance.

Introduction and History

The United Nations Conference on Environment and Development

(UNCED), also known as the Rio Summit, or Rio Conference, Earth Summit (Portuguese: Eco '92) was a major United Nations conference held in Rio de Janeiro, Brazil from 3 to 14 June, 1992.

In the conference 172 governments participated with 108 sending their heads of state or government. There are some 2,400 representatives of non-governmental organizations (NGOs) attended with 17,000 people at the parallel NGO "Global Forum", who had Consultative Status.

Rio produced two international agreements, two statements of principles and a major action agenda on worldwide sustainable development.

The Earth Summit resulted in the following five documents :

1. *Rio Declaration on Environment and Development*—It resulted into 27 principles define the rights and responsibilities of nations as they pursue human development and well-being.
2. *Forest Principles*—A statement of principles to guide the management, conservation and sustainable development of all types of forests, which are essential to economic development and the maintenance of all forms of life.
3. *Convention on Biological Diversity*—The Convention on Biological Diversity requires that countries adopt ways and means to conserve the variety of living species, and ensure that the benefits from using biological diversity are equitably shared.
4. *Framework Convention on Climate Change (UNFCCC)*—The aim of the United Nations Framework Convention on Climate Change is to stabilize greenhouse gases in the atmosphere at levels that will not dangerously upset the global climate system. This will require a reduction in our emissions of such gases as carbon dioxide, a by-product of the use of burning fuels for energy.
5. *Agenda 21*—It is a blueprint on how to make development socially, economically and environmentally sustainable.

Agenda 21 is a programme run by the United Nations (UN) related to sustainable development. It is a comprehensive blueprint of action to be taken globally, nationally and locally by organizations of the

UN, governments, and major groups in every area in which humans directly affect the environment.

Implementation of Agenda 21 by member-states remains essentially voluntary.

There are 40 chapters in the Agenda 21, divided into four main sections

Section I: Social and Economic Dimensions

It includes combating poverty, changing consumption patterns, population and demographic dynamics, promoting health, promoting sustainable settlement patterns, integrating environment and development into decision making.

Section II: Conservation and Management of Resources for Development

It includes atmospheric protection, combating deforestation, protecting fragile environments, conservation of biological diversity (biodiversity), and control of pollution.

Section III: Strengthening the Role of Major Groups

It includes the roles of children, youth, women, NGOs, local authorities, business and workers.

Section IV: Means of Implementation

Implementation includes science, technology transfer, education, international institutions and mechanisms and financial mechanisms.

It is striking to note that most of the issues forming integral part of Agenda 21 are already brought forth by Mahatma Gandhi a century back when there was absolutely no pollution. All the problems were envisaged long ago and given probable solution to it. There may be repetition of certain points and quotations at more than one place because of their relevance.

In this article we will concentrate on two chapters that is Chapter 34 and 36 from Section IV of Agenda 21 and compare the same with Gandhian philosophy as follows:

1. Transfer of Environmentally Sound Technology, Cooperation and Capacity-Building—Chapter 34

Environmentally sound technologies protect the environment, are less polluting, use all resources in a more sustainable manner, recycle more of their wastes and products, and handle residual wastes in a more acceptable manner than the technologies for which they were substitutes.

Environmentally sound technologies are those which generate low or no waste, for the prevention of pollution. They also cover technologies for treatment of pollution if generated.

Environmentally sound technologies are not just individual technologies, but total systems which include know-how, procedures, goods and services, and equipment as well as organizational and managerial procedures. This implies that when discussing transfer of technologies, the human resource development and local capacity-building aspects of technology choices, including gender-relevant aspects, should also be addressed. Environmentally sound technologies should be compatible with nationally determined socio-economic, cultural and environmental priorities.

There is a need for favourable access to and transfer of environmentally sound technologies, in particular to developing countries, through supportive measures that promote technology cooperation and that should enable transfer of necessary technological know-how as well as building up of economic, technical, and managerial capabilities for the efficient use and further development of transferred technology. Technology cooperation involves joint efforts by enterprises and governments, both suppliers of technology and its recipients. Therefore, such cooperation entails an iterative process involving government, the private sector, and research and development facilities to ensure the best possible results from transfer of technology. Successful long-term partnerships in technology cooperation necessarily require continuing systematic training and capacity-building at all levels over an extended period of time.

The activities proposed in this chapter aim at improving conditions and processes on information, access to and transfer of technology (including the state-of-the-art technology and related know-how), in particular to developing countries, as well as on capacity-building and cooperative arrangements and partnerships in the field of technology, in order to promote sustainable development. New and efficient technologies will be essential to increase the capabilities, in particular of developing countries, to achieve sustainable development, sustain the world's economy, protect the environment, and alleviate poverty and human suffering. Inherent in these activities is the need to address the improvement of technology currently used and its replacement, when appropriate, with more accessible and more environmentally sound technology.

Gandhian Views

According to Gandhiji the industrialisation and urbanisation have created multiple problems and miseries for the modern man. In the process, we have tried to explain in detail as to how industrialisation, modern civilisation and rapid urbanisation have created chaos, lop-sided development, rapid depletion of natural resources and danger for natural environment.

Gandhi was critical of modern civilisation, rapid industrialisation and galloping urbanisation. In this context, we find the pertinence of Gandhian ideas.

Long before the modern environmentalists, Gandhi correctly realised that rapid industrialisation cannot be the panacea to all ills. Increasing industrialisation in today's world has not reduced social inequalities, but has rather resulted in further differentiations. Increasing use of technology has led to greater heterogeneity, greater inequalities and greater unaltruistically oriented behaviour.

Gandhi regarded industrialisation detrimental to growth of a non-violent and eco-friendly society. In his ideal society, as in the classical anarchist model, there would be complete decentralisation of political and economic system and self-sufficient, barter type of village economy would be the desired model.

Machinery has, in his judgement; three-essential attributes. First, it can be duplicated or copied. Secondly, there is no limit to its growth or evolution. Thirdly, it appears to possess a will or genius of its own that operates as the inevitable law of displacement of the labour. Once the machine is created and allowed to operate, it goes more and more out of human control.

Ideally, Gandhi regards all machinery as thoroughly undesirable. Once he commented: "Today machinery merely helps a few to ride on the backs of millions. The impetus behind it all is not philanthropy to save labour but greed. It is against this constitution of things that I am fighting with all my might".

His arguments against machinery can broadly be divided into two categories: ethical and economic. The arguments of the first category run as follows:

(i) Labour is a value relative to non-violence and machinery tends to undermine it.
(ii) Machines are repugnant in his thinking to the good life.
(iii) The invention of machinery has led to the growth of the factory system which has reduced the masses of men to the condition of slaves.
(iv) The technological advancement has led to the growth of the monetary exchange system which is characterised by inequality and exploitation.
(v) Machinery has led to the growth of economic competition which undermines the process of cooperation.

The arguments falling in the second category are:

The displacement of human labour is an essential characteristic of a machine and a great argument against it, introductions of machines results in employment of a few and unemployment of many, it saves labour and provides leisure, leisure results in wastage of time, and potential cause of demoralisation.

Machines lead to the concentration of wealth in the hands of a few.

Machinery inevitably leads to mass production and mass production necessarily leads to over production, storage, transportation. Hence, Gandhiji used to say that instead of mass production there should be production by masses.

The application of machinery in agriculture may destroy the fertility of the soil and lead to loss of production.

Machines lead to growth of congested, unhygienic cities, speed of travel, etc., which results in the loss of health.

Gandhi felt that the present industrialisation and use of large scale machinery was not very healthy and resulted in serious economic dislocation. Dead machinery must not be pitted against millions of living machines. As Gandhi once commented :

"Mechanisation is good when the hands are too few for the work intended to be accomplished. It is an evil when there are more hands than required for the work, as in India."[4]

Large scale industrialisation perpetuates war and many other evils and all the naturalness come to an end.

In conclusion, Gandhi rejected the modern industrial-urban concept of development for its anti-democratic, anti-humanitarian, and exploitative features. In its place Gandhi offers the ideal of the economically self-sufficient, politically self-governing and culturally non-violent village republic as the guarantee of genuine democracy, true humanism, civilising non-violence and lasting peace. Thus Gandhiji was in favour of technology and development of cottage and small scale industries at village level because these industries are localised, energy saver, job intensive and less polluting. According to him cities should be as store and forwarding houses and no production in cities to prevent congestion and pollution.

2. *Promoting Education, Public Awareness and Training—Chapter 36*

Education, raising of public awareness and training are linked to virtually all areas in Agenda 21. This chapter sets out broad proposals, while specific suggestions related to sector-wise issues are contained in other chapters. Programme areas described in the present chapter are:

(a) Reorienting education towards sustainable development;
(b) Increasing public awareness; and
(c) Promoting training.

Education, including formal education, public awareness and training should be recognized as a process by which human beings and societies can reach their fullest potential. Education is critical for promoting sustainable development and improving the capacity of the people to address environment and development issues. The basic education provides the underpinning for any environmental and development education, the latter needs to incorporate as an essential part of learning. Both formal and non-formal education is indispensable to changing people's attitudes so that they have the capacity to assess and address their sustainable development concerns. It is also critical for achieving environmental and ethical awareness, values and attitudes, skills and behaviour consistent with sustainable development and for effective public participation in decision-making.

To be effective, environment and development education should deal with the dynamics of both the physical/biological and socio-economic environment and human (which may include spiritual) development, should be integrated in all disciplines, and should employ formal and non-formal methods and effective means of communication.

Gandhian Views: According to Gandhiji education should include element of morality and ethics to motivate individual to perform his duty better. Gandhiji was of the opinion that hand, head and heart to be simultaneously trained. He also strongly promoted physical education which he would prepare the youth in earning his livelihood by bread labour. He was firm believer that education should be linked with training and production that is vocational and job oriented training. Intellectual training or training of head was to be done through literature and social science. Mind is trained by observations and experience and experiments. Training of mind is necessary because mind wearies and keeps on wandering and body also demands more and more and must be trained to curb the wants. Hence poor and rich both are not happy.

The ethical people are better human beings. They are qualitatively very superior like saints. Gandhi's man is not economical man but ethical man because such man is fit for sustainable society. Self regulated individual could contribute better to a sustainable society because there would be least opportunity to govern. Gandhiji always stood for 'Minimum State'. Minimum state is only possible when individual is self regulated and self disciplined.

The education that has no relation to actual living conditions is not national education essential for democracy to function in fruitful ways. The education to be proper must help shape lives and answer the wants of people. It is no education if it fails to make farmer's son a better farmer. Present primary education is of no use in after-life becomes useless. Similarly higher education makes us foreigner in our own country, higher he goes further he moves from the house. Gandhiji says that "by education it means all round drawing out of the best in the child and man.

On this basis Gandhi announced new system of education that is called as Nai Talim. New education branches into three stages. First stage up to the age of eight, children is not taught reading or writing. It consists of manual training under supervision of an educationist Education should appear as a play. Second stage is from nine to sixteen. This is primary education, the basic education covering standard matriculation subjects plus vocational subjects except English. Third stage lasts after nine years after primary education. During this period youngster will have education according to his wishes and circumstances Education should increase all types of subject history, geography, arithmetic, science, languages—vernacular and Hindi, painting, music, handicraft. Education should be self supporting and self confident and courageous.

Gandhiji felt that agriculture taught should be intimately related to rural conditions. College engineering students should be attached to different industries paying for the training of the graduates. Thus the textile mills associations should be under the supervision of the state. Medical colleges should be attached to certified hospitals. Agriculture College should part with extensive agriculture experience. Gandhian philosophy continues to hold relevance and rural institutes inspirited in their establishment by Gandhian philosophy have a special responsibility to establish the intrinsic connection between the classroom the community surrounding it.

Conclusion

Nature has its own self-cleaning system. But it has limitation beyond which it cannot work. This system has been adversely affected due to manmade activities which have led to degradation of environment through pollution of air, water and land and exploitation of natural resources.

Around a century ago environment was clean, virgin, pure, uncontaminated and acceptable to all. Now it is unclean, impure, contaminated and unacceptable to living beings. The worldwide environmental problem/degradation is well felt. This is due to following reasons :

1. Industrialisation;
2. Urbanisation;
3. Population explosion;
4. Throw away concepts of human beings; and
5. Excessive exploitation of Natural Resources.

Besides this due to greedy nature of the people there is heavy reduction in Natural Resources (both renewable and non-renewable) that is air, water, minerals, fuel—petrol, diesel, kerosene, gases, wood etc. (some of them are renewed after long period of time) and none of these resources are free of cost. There is famous quoting by Mahatma Gandhi—"we have enough to satisfy all people's need but not enough to satisfy some people's greed". In New Delhi Declaration views similar to above have been expressed as "Some for all rather than more for some".

In rural area generation of energy is at most important to run rural industries and other activities. These industries are energy savers, less wasteful and less damaging to environment and are essential to supplement agricultural growth.

Moreover, increasing global warming (effect of Carbon dioxide), the hole in stratosphere of ozone layer in the atmosphere of sky, and depletion in biodiversity has threatened the human, animal, plant life and aesthetics of property.

All above problems could be solved by studying, accepting and following Gandhian philosophy and practices. This will also save future generation suffering from psychosomatic diseases (physical and mental) problems.

The environmental problems were envisaged by Mahatma Gandhi a century ago when actually there was no environmental problem and hence Gandhiji was real visionary. According to Gandhian philosophy problem of environmental degradation is in the mind of individual. He should change himself from inside out for which individual must be spiritual and religious. Hence, prayer was made part of daily routine in Gandhi *ashram*. The root cause is human greed, needs and wants. This gives rise to vicious circle that is consumerism, industrialisation, mass production, huge requirement of raw material and fuel, large storage space, heavy transportation and finish product in bulk quantity. The technology and machinery required will displace labour due to automation. This will give rise to another vicious circle that is displacement of labour, unemployment, poverty, environmental degradation. These two vicious circles will have negative effect on environment. Hence, concept of *swaraj* was put forth by Gandhiji that is total liberation, self reliance and self sufficiency. Hence, Gandhiji gave slogan for youth "to go back to villages". He insisted for village and handicraft industries which are less machine dependent, labour intensive, energy saver and protects environment. Hence, khadi industries were promoted.

This Gandhian approach of protection of environment and reduction of pollution was holistic socially and economically.

REFERENCES

1. *http://en.wikipedia.org/wiki/Agenda_21*
2. *http://habitat.igc.org/agenda21/*
3. Prabhath, S.V. (Ed). (2010), *Gandhi Today*, Serials Publications, New Delhi.
4. *Gandhi M.K., Quoted by Mathurs (Ed.), ETMG*, p. 476.
5. *Ibid.*, pp. 249.
6. Kumar, S.B., Environmental Problems and Gandhian Solutions, The only ray of hope to the present ailing world, Deep & Deep Publications (Pvt.) Ltd., New Delhi.
7. Mahatma Gandhi, *Last Phase, Vol. II* (1958), p. 65.
8. Man v. Machine (online) cited 6 March 2002 Available from <url:www.mkgandhi.org.>

Index